Algorithms and Combinatorics 22

Springer
Berlin
Heidelberg
New York
Barcelona
Hong Kong
London
Milan
Paris
Singapore
Tokyo

Joel Spencer

The Strange Logic of Random Graphs

With 13 Figures

Springer

Joel Spencer
Courant Institute
New York University
251 Mercer Street
New York, NY 10012-1110
USA
e-mail: spencer@cs.nyu.edu

Library of Congress Cataloging-in-Publication Data applied for
Die Deutsche Bibliothek – CIP-Einheitsaufnahme
Spencer, Joël:
The strange logic of random graphs / Joel Spencer. – Berlin; Heidelberg; New York; Barcelona; Hong Kong; London; Milan; Paris; Singapore; Tokyo: Springer, 2001
(Algorithms and combinatorics; 22)
ISBN 3-540-41654-4

Mathematics Subject Classification (2000): 05C80, 03B10

ISSN 0937-5511
ISBN 3-540-41654-4 Springer-Verlag Berlin Heidelberg New York

Springer-Verlag Berlin Heidelberg New York
a member of BertelsmannSpringer Science+Business Media GmbH
http://www.springer.de

Printed in Germany

Typeset in LaTeX by the author. Edited and reformatted by Johannes Küster, typoma, www.typoma.com, using a Springer LaTeX macro package.
Printed on acid-free paper SPIN 10774693 46/3142LK - 5 4 3 2 1 0

To MaryAnn

Contents

Part I. Beginnings

Part II. Random Graphs

Part I

Beginnings

0. Two Starting Examples

The core of our work is the study of Zero-One Laws for sparse random graphs. We may think of this study as having two sources. The first is the Zero-One Law for random graphs with constant probability, as given in Section 0.1. The second is the notion of evolution of random graph as discussed in Section 1.1.1. In that evolution it is central that the edge probability p be taken not just as a constant but as a function $p = p(n)$ of the total number of vertices. In Section 0.2 we examine such an evolution in the much easier case of a random unary predicate. To allow an easy introduction we avoid a plethora of notation in this chapter, the technical preliminaries – including many key definitions – are left for Chapter 1 and beyond.

0.1 A Blend of Probability, Logic and Combinatorics

We will be looking at labelled graphs G on n vertices. For convenience we'll call the vertices $\{1, \ldots, n\}$. The number of such graphs is precisely $2^{\binom{n}{2}}$ as each of the $\binom{n}{2}$ pairs i, j can be either adjacent or not adjacent. Consider a graph property – for example, the property of containing a triangle. Call the property A. We set $\mu_n(A)$ equal the proportion of labelled graphs on n vertices that have the property A. A precise evaluation of $\mu_n(A)$ might be very difficult. We start slowly.

Claim 0.1.1 $\lim_{n\to\infty} \mu_n(A) = 1$

Rather than proportion, it will be easier to work [throughout this book] with probabilities. Imagine that every pair i, j of vertices flips a fair coin to decide whether or not to be adjacent. We call the outcome the random graph $G(n, \frac{1}{2})$, which is defined in Section 1.1. We can and shall interpret $\mu_n(A)$ as the probability that this random graph has property A.

With this interpretation we give a simple argument (one of many) for Claim 0.1.1. Split the vertices into $s = \lfloor n/3 \rfloor$ disjoint triples. A triple i, j, k forms a triangle with probability precisely $\frac{1}{8}$. These are independent events as they involve distinct coin flips. Thus the probability that none of the s triples form a triangle is $(7/8)^s$. This goes to zero as n, and therefore s, goes to infinity. But this is an upper estimate (in some sense a very poor one but it suffices for our purposes) of the probability that there is no triangle whatsoever. When a nonnegative sequence is bounded from above by a sequence going to zero it must itself go to zero. So the probability that there is no triangle goes to zero.

Definition 0.1 *When* $\lim_{n\to\infty} \mu_n(A) = 1$ *holds we say that property* A *occurs* almost surely, *or, equivalently, that almost all graphs have property* A. *When* $\lim_{n\to\infty} \mu_n(A) = 0$ *we say that property* A *holds* almost never, *or, equivalently, that almost no graphs have property* A.

We note that this notation is not standard. A number of authors use the term *asymptotically almost surely* for the above concept and reserve almost surely for events that have probability one.

Let's consider, without proofs, some other examples of properties A. Almost all graphs are connected. Almost no graphs are planar. Almost no graphs have an isolated vertex. Almost all graphs have an induced pentagon. Is there a strict dichotomy (what we'll later call a Zero-One Law) between almost all and almost no? Of course not. The average graph will have $\frac{1}{2}\binom{n}{2}$ edges. Let A be the event that the graph G has more than $\frac{1}{2}\binom{n}{2}$ edges. It is not difficult to show $\lim_{n\to\infty} \mu_n(A) = \frac{1}{2}$, that asymptotically half the graphs have more than $\frac{1}{2}\binom{n}{2}$ edges. Sometimes a "silly" example can be instructive. Let A be the event that n itself is even. Then $\mu_n(A)$ is one when n is even and zero when n is odd, we aren't even looking at the graph. Here $\mu_n(A)$ does not approach a limit as $n \to \infty$! Still, these properties that avoid the strict dichotomy are somewhat suspect, the earlier properties have much more of a naturalness to them.

We would like to say that natural properties hold either almost surely or almost never. But what properties shall we call natural? For most of this book we shall deal with *first order properties*, as defined in Section 1.2. This is a notion long studied by logicians. How well it captures "naturalness" is discussed in Section 8.1.3 with some less than positive comments but our reason for using it is quite pragmatic: we can prove something remarkable.

Theorem 0.1.2 (Fagin-GKLT). *Let* A *be* any *first order property. Then*

$$\lim_{n\to\infty} \mu_n(A) = 0 \text{ or } 1$$

That is, every first order sentence holds either almost surely or almost never.

GKLT refers to Glebskii, Kogan, Liagonkii and Talanov [8]. The proof of this theorem we give here, basically from Fagin [7], is a blend of combinatorics, probability and logic. For every pair of nonnegative integers r, s we define a particular property of special importance.

Definition 0.2 *The* r, s *extension statement, denoted* $A_{r,s}$, *is that for all distinct vertices* $x_1, \ldots, x_r$ *and* $y_1, \ldots, y_s$ *there exists a vertex* z *distinct from them all which is adjacent to all of the* $x_1, \ldots, x_r$ *and to none of the* $y_1, \ldots, y_s$.

We naturally include the cases $r = 0$ and $s = 0$ so that, for example, $A_{1,0}$ is that every vertex has a neighbor. When z is adjacent to all the x's and none of the y's we call z a *witness*, a term that will appear in later contexts as well. The probability part of the proof consists of showing that

Claim 0.1.3 *For all $r, s \geq 0$ the r, s extension statement $A_{r,s}$ holds almost surely.*

For a given $x_1, \ldots, x_r$ and $y_1, \ldots, y_s$ let $\textsc{Noz}[x_1, \ldots, y_s]$ be the event that there is no witness z. (The notation $\textsc{Noz}$ is meant to suggest "no z.")

Claim 0.1.4 $\Pr[\textsc{Noz}] = (1 - 2^{-r-s})^{n-r-s}$.

Proof: There are $n - r - s$ potential witnesses z's. Each has probability 2^{-r-s} of being a witness, as $r + s$ coin tosses must come up in a particular way. But the events "z is not a witness" are mutually independent over the z's as they involve disjoint sets of coin tosses. Thus the probability the no z is a witness is $(1 - 2^{-r-s})^{n-r-s}$.

While $\Pr[\textsc{Noz}] \rightarrow 0$ that by itself only shows that for a particular $x_1, \ldots, x_r$ and $y_1, \ldots, y_s$ there is almost surely a witness z. The event $A_{r,s}$ is logically equivalent to saying $\textsc{Noz}$ fails for all choices of x's and y's. Turning things around, the event $\neg A_{r,s}$ is the disjunction of the events $\textsc{Noz}$ over all possible $x_1, \ldots, x_r$ and $y_1, \ldots, y_s$. There are a total of $\binom{n}{r}\binom{n-r}{s}$ choices for the x's and y's.

Now we need an absolutely elementary fact: The probability of the disjunction of events is *at most* the sum of the probabilities of the events. Equality occurs only when the events are disjoint, which will not be the case here. The actual calculation of the probability of a disjunction can be quite complicated (involving, e.g., the Inclusion-Exclusion laws) but it is surprising how often the above fact will suffice for our purposes.

In applying our fact all the events have the same probability so the sum is actually a product and

$$\Pr[\neg A_{r,s}] \leq \binom{n}{r}\binom{n-r}{s}(1 - 2^{-r-s})^{n-r-s} \tag{1}$$

Now, recalling r, s are fixed, we take the limit of the right hand side of 1 as $n \rightarrow \infty$. The term $\binom{n}{r}\binom{n-r}{s}$ is a polynomial in n. The term $(1-2^{-r-s})^{n-r-s}$ is an exponential in n, going to zero as $1 - 2^{-r-s} < 1$. Polynomial growth times exponential decay goes to zero. We've bounded $\Pr[\neg A_{r,s}]$ from above by a function going to zero and hence $\neg A_{r,s}$ holds almost never. But then $A_{r,s}$ hold almost surely, giving Claim 0.1.3.

Now our argument makes a surprising turn into the infinite.

Definition 0.3 *A graph G is said to have the* Alice's Restaurant *property if it satisfies the r, s extension statement $A_{r,s}$ for all nonnegative integers r, s. Equivalently: if for all pairs of disjoint finite sets X, Y of vertices there* *exists z not in their union which is adjacent to all $x \in X$ and no $y \in Y$. Equivalently: given any finite set X of size, say, s there are witnesses $z \notin X$ with all 2^s possible adjacency patterns to X.*

This colorful term was first used by Peter Winkler. It refers to a popular song by Arlo Guthrie whose refrain – you can get anything you want at Alice's Restaurant – captures the spirit of the property. No finite graph G can have the Alice's Restaurant property since one could take X to be the entire vertex set and $Y = \emptyset$ and then there would be no witness z. The surprise comes when we look at countable graphs.

Theorem 0.1.5. *There is a unique (up to isomorphism) countable graph G satisfying the Alice's Restaurant property.*

Proof of Uniqueness: Let G_1, G_2 be countable graphs satisfying the Alice's Restaurant property, label their vertices $a_1, a_2, \ldots$ and $b_1, b_2, \ldots$ respectively. We will find a bijection $\phi\colon G_1 \to G_2$ in an infinite number of stages, which shall alternate between LeftStep and RightStep, beginning with a LeftStep. At the beginning ϕ is nowhere defined. At each step we will define one more value of ϕ. (While the first (left) step can be considered part of the general procedure below we note it always consists of setting $\phi(a_1) = b_1$.) Say that after s steps we have defined $\phi(x_i) = y_i$ for $1 \le i \le s$. We shall require inductively that ϕ is an isomorphism between its domain and range, i.e., that x_i, x_j are adjacent if and only if y_i, y_j are adjacent.

We define a RightStep. Let y_{s+1} be the first vertex of G_2 (by which we mean that vertex with the smallest index when written b_j) which is not one of the $y_1, \ldots, y_s$. We define, and this is the critical point, x_{s+1} to be the first vertex of G_1 which is not one of the $x_1, \ldots x_s$ and so that defining $\phi(x_{s+1}) = y_{s+1}$ retains the inductive property – i.e., that x_{s+1} is adjacent to x_i (with $1 \le i \le s$) if and only if y_{s+1} is adjacent to y_i. We are looking for an x_{s+1} with a particular set of adjacencies to the $x_1, \ldots, x_s$. The existence of such an x_{s+1} follows from the Alice's Restaurant property of the graph G_1.

A LeftStep is similar, taking x_{s+1} to be the first vertex of G_1 which is not one of the $x_1, \ldots, x_s$ and then y_{s+1} to be the first vertex of G_2 which is not one of the $y_1, \ldots, y_s$ which is adjacent to y_i (with $1 \le i \le s$) if and only if x_{s+1} is adjacent to x_i. The Alice's Restaurant for the graph G_2 guarantees the existence of y_{s+1} and we set $\phi(x_{s+1}) = y_{s+1}$.

The final ϕ obtained by this procedure will be an isomorphism between its domain and range. But any vertex in G_1 has some label, say a_u, and so will be in the domain after at most u LeftSteps, since at each LeftStep the least vertex of G_1 not already taken is placed in the domain. Similarly, any vertex in G_2 has some label, say b_v, and so will be in the range after at most v RightSteps, since at each RightStep the least vertex of G_2 not already taken is placed in the range. Thus ϕ is a bijection from G_1 to G_2 which preserves adjacency and hence G_1, G_2 are isomorphic.

An Example: In the partial picture below we first set $\phi(a_1) = b_1$. The next step is a RightStep and b_2 is the first unused vertex of G_2. It is adjacent to b_1. The first unused vertex of G_1 adjacent to a_1 is a_4 so we set $\phi(a_4) = b_2$. The next step is a LeftStep and a_2 is the first unused vertex of G_1. It is adjacent to a_1 and not a_4 so we seek an unused vertex of G_2 adjacent to b_1 and not b_2, the first one is b_6 and we set $\phi(a_2) = b_6$. The next step is a RightStep and b_3 is the first unused vertex of G_2. It is adjacent to b_2 and not to b_1 nor b_6 so we seek an unused vertex of G_1 adjacent to a_4 and not to a_1 nor a_2. The Alice's Restaurant property of G_1 assures us that such a vertex exists, if the first one is a_{17} we set $\phi(a_{17}) = b_3$ and continue.

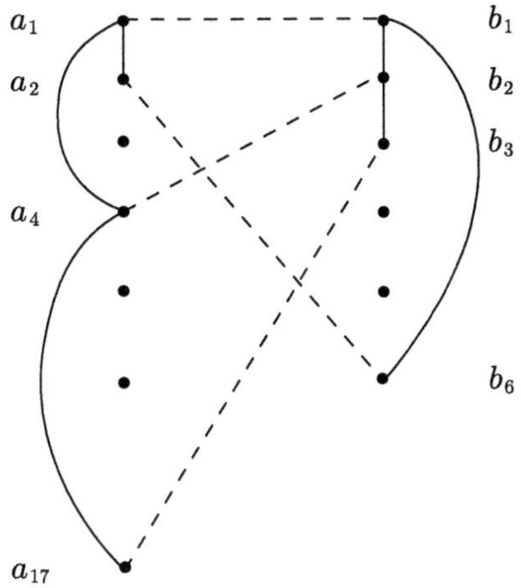

Fig. 0.1.

Existence (Proof 1): Let the vertices be $0, 1, 2, \ldots$. For $s \geq 1$ let the 2^s vertices $2^s \leq i < 2^{s+1}$ have all possible adjacency patterns with $0, \ldots, s-1$. Explicitly, when $2^s \leq i < 2^{s+1}$ write $i = 2^s + \sum_{j=0}^{s-1} \epsilon_j 2^j$. For $0 \leq j < s$ have i, j adjacent if and only if $\epsilon_j = 1$. (Not all adjacencies have been specified by this procedure, those that have not can be filled in arbitrarily.) Any finite set X has a maximal value s and so there will be a witness $i \in [2^s, 2^{s+1})$ with any desired adjacency pattern to X.

Existence (Proof 2): Start with a countable vertex set on which no adjacencies have been determined. Make a countable list (X_i, Y_i) of pairs of disjoint finite sets from the vertex set. At step i take a vertex z_i not previously used (not in $X_1, \ldots, X_i$, $Y_1, \ldots, Y_i$ nor $z_1, \ldots, z_{i-1}$) and make it adjacent to all of X_i and none of Y_i. At the end of the countable procedure some pairs have not had their adjacency determined, they can be set arbitrarily. But any finite pair (X, Y) appeared in the countable list as some position i and so has its witness $z = z_i$.

Existence (Proof 3): Since each $A_{r,s}$ holds almost surely the theory T generated by them is consistent and hence has a countable model, as discussed more generally in Sections 1.5 and 1.6.

The final portion of Theorem 0.1.2 uses Logic. Consider the theory T of graphs with the sentences $A_{r,s}$. (That is, add the $A_{r,s}$ as axioms.) This theory has no finite models (that is, graphs satisfying the axioms) and has a unique (up to isomorphism) countable model. From the Gödel Completeness Theorem, a basic but very deep result in logic, we deduce (as described in more detail in Section 1.5) that the theory T is complete. This means that for any sentence B either B or $\neg B$ is deducible in the theory – provable from the axioms $A_{r,s}$.

Suppose B is provable in T. Proof is finite and so there is a proof using only finitely many of the axioms, call them A^i for $1 \leq i \leq u$, each being of the form $A_{r,s}$. (This reduction to a finite number of axioms is critical and sometimes referred to as the Compactness Principle.) Any G that satisfies the conjunction $\wedge_i A^i$ must satisfy B. Complementing, any G satisfying $\neg B$ must satisfy the disjunction $\vee_i \neg A^i$. The probability of a disjunction is at most the sum of the probabilities and so for any n

$$\mu_n(\neg B) \leq \sum_{i=1}^{s} \mu_n(\neg A^i)$$

The limit of a *finite* sum of sequences is the sum of their limits so

$$\lim_{n\to\infty} \sum_{i=1}^{s} \mu_n(\neg A^i) = \sum_{i=1}^{s} \lim_{n\to\infty} \mu_n(\neg A^i) = \sum_{i=1}^{s} 0 = 0$$

Therefore $\neg B$ holds almost never. Therefore B holds almost surely.

The only other case is when $\neg B$ is provable in T. The roles of B and $\neg B$ are now reversed. We deduce B holds almost never. So either B holds almost surely or almost never, completing the proof of Theorem 0.1.2.

The Fagin-GKLT Theorem 0.1.2 deals with asymptotics but speaks only about finite graphs, infinite graphs never appear in the statement. Yet this proof involves "going to the infinite and coming back". It was that aspect that first convinced this author (among many) of the beauties of the subject. A rough analogy can be made to the use of the complex numbers to prove statements about the reals. Mathematics works in strange ways. We shall explore a number of techniques that lead to Zero-One Laws but the use of infinite graphs shall remain a strong motif throughout this work.

0.2 A Random Unary Predicate

We turn now away from graphs to a rather easier random model which illustrates many of the concepts we shall deal with. We call it the simple unary predicate with parameters n, p and denote it by $\mathrm{SU}(n, p)$. The model is over a universe Ω of size n, a positive integer. We imagine each $x \in \Omega$ flipping a coin to decide if $U(x)$ holds, and the coin comes up heads with probability p. Here we have p real, $0 \le p \le 1$. Formally we have a probability space on the possible U over Ω defined by the properties $\Pr[U(x)] = p$ for all $x \in \Omega$ and the events $U(x)$ being mutually independent. We consider sentences in the first order language. In this language we have only equality (we shall always assume we have equality) and the unary predicate U. The cognescenti should note that Ω has no further structure and in particular is not considered an ordered set as in Section 10.7.

This is a spartan language. One thing we can say is

$$\text{YES} := \exists_x U(x),$$

that U holds for some $x \in \Omega$. Simple probability gives

$$\Pr[\text{SU}(n,p) \models \text{YES}] = 1 - (1-p)^n$$

As p moves from zero to one $\Pr[\text{YES}]$ moves monotonically from zero to one. We are interested in the asymptotics as $n \to \infty$. At first blush this seems trivial: for $p = 0$, $\text{SU}(n,p)$ never models YES while for any constant $p > 0$,

$$\lim_{n\to\infty} \Pr[\text{SU}(n,p) \models \text{YES}] = \lim_{n\to\infty} 1 - (1-p)^n = 1$$

In an asymptotic sense YES has already almost surely occurred by the time p reaches any positive constant.

This leads us to a critical notion. *We do not restrict ourselves to p constant but rather consider $p = p(n)$ as a function of n.* What is the parametrization $p = p(n)$ that best enables us to see the transformation of $\Pr[\text{SU}(n,p(n)) \models \text{YES}]$ from zero to one. Some reflection leads to the parametrization $p(n) = c/n$. If c is a positive constant then

$$\lim_{n\to\infty} \Pr[\text{SU}(n,p(n)) \models \text{YES}] = \lim_{n\to\infty} 1 - \left(1 - \frac{c}{n}\right)^n = 1 - e^{-c}$$

(Technically, as $p \le 1$ always, this parametrization is not allowable for $n < c$ – but since our interest is only with limits as $n \to \infty$ this will not concern us.) If we think of c going from zero to infinity then the limit probability is going from zero to one. We shall not look at the actual limits here but only in whether the limits are zero or one.

Repeating the notation of Section 0.1 we say that a property A holds *almost always* if $\lim_{n\to\infty} \Pr[\text{SU}(n,p(n)) \models A] = 1$. We say that A holds *almost never* if the above limit is zero or, equivalently, if $\neg A$ holds almost surely. Note, however, that these notions depend on the particular function $p(n)$. This notion is extremely general. Whenever we have for all sufficiently large positive integers n a probability space over models of size n then we can speak of a property A holding almost surely or almost never. For the particular property YES the exact results above have the following simple consequences:

• If $p(n) \ll n^{-1}$ then YES holds almost never.
• If $p(n) \gg n^{-1}$ then YES holds almost surely.
Thus, for example, when $p(n) = n^{-1.01}$ YES holds almost never while when $p(n) = n^{-0.99}$ YES holds almost surely.

We shall say n^{-1} is a *threshold function* for the property YES. More generally, suppose we have a notion of a random model on n vertices with probability p of some predicate. We say $p_0(n)$ is a threshold function for a property A if whenever $p(n) \ll p_0(n)$ the property A holds almost never and whenever $p(n) \gg p_0(n)$ then the property A holds almost surely. This notion, due to Paul Erdős and Alfred Rényi, says roughly that $p_0(n)$ is the "region" around which $\Pr[A]$ is moving from near zero to near one. We'll have much more to say about the evolution of random *graphs* in Section 1.1.1 and elsewhere. The threshold function, when it exists, is not totally determined – we could have taken $5/n$ as the threshold function for YES – but is basically determined up to constant factors. In a rough way we think of $p(n)$ increasing through the functions of n – e. g. from n^{-2} to n^{-1} to $n^{-1}\ln n$ to $\ln^{-5} n$ – and the threshold function is that place where $\Pr[A]$ changes.

A natural problem for probabilists is to determine the threshold function, if one exists, for a given property A. For logicians the natural question would be to determine all possible threshold functions for all properties A expressible in a given language L. Unfortunately there are technical difficulties (especially with later more complex models) with threshold functions – properties A need not be monotone, threshold functions need not exist. Worst of all, as shown in Section 8.3.4, the limits of probabilities might not exist. Rather, the logician looks for a *Zero-One Law* of which the following is prototypical:

Theorem 0.2.1. *Let $p = p(n)$ satisfy $p(n) \gg n^{-1}$ and $1 - p(n) \gg n^{-1}$. Then for* any *first order property A*

$$\lim_{n\to\infty} \Pr\big[\mathrm{SU}(n,p) \models A\big] = 0 \text{ or } 1$$

Further, the limiting value depends only on A and not on the choice of $p(n)$ within that range.

Our approach to this theorem is to find an explicit theory T such that
• Every $A \in T$ holds almost surely
• T is complete
Will this suffice? When $T \models B$ finiteness of proof gives that B follows from some $A_1, \ldots, A_s \in T$ and hence from $A_1 \wedge \ldots \wedge A_s$. But the finite conjunction of events holding almost surely holds almost surely so B would hold almost surely. By completeness, either $T \models B$ or $T \models \neg B$, and in the latter case $\neg B$ holds almost surely so that B holds almost never.

In our situation T is given by two simple schema.

1. (For $r \geq 1$) There exist distinct $x_1, \ldots, x_r$ with $U(x_i)$ for $1 \leq i \leq r$.
2. (For $r \geq 1$) There exist distinct $x_1, \ldots, x_r$ with $\neg U(x_i)$ for $1 \leq i \leq r$.

Note that the number X of x with $U(x)$ has Binomial Distribution with parameters $n, p(n)$ – that the event $X \geq r$ holds almost surely follows from basic probabilistic ideas from the assumption $np(n) \to \infty$. The second schema follows from $n\left(1 - p(n)\right) \to \infty$, reversing the roles of U and $\neg U$.

Why is this T complete? Proving completeness of a theory T is bread and butter to the logic community – from the myriad of methods we choose a combinatorial approach based on the Ehrenfeucht game, as described in Chapter 2. Let $t \geq 1$ be arbitrary and let M_1, M_2 be two countable models of T. It suffices to show that Duplicator wins the game $\text{EHR}(M_1, M_2; t)$. Here, however, our models have only a unary predicate and no edges. Thus Duplicator's chore is relatively simple. When Spoiler selects $v \in M_1$ (say) with $U(v)$ Duplicator must select $v' \in M_2$ with $U(v')$. Similarly, when Spoiler selects $v \in M_1$ (say) with $\neg U(v)$ Duplicator must select $v' \in M_2$ with $\neg U(v')$. Further, Duplicator cannot select a vertex already selected unless Spoiler has done so.

In our case the Duplicator strategy is simple. A model M of T must have an infinite number of $x \in M$ with $U(x)$ (as for all $r \geq 1$ it must have at least r such x) and, similarly, an infinite number of $x \in M$ with $\neg U(x)$. Now when Spoiler selects, say, a new $x \in M_1$ with $U(x)$ Duplicator simply selects a new $x' \in M_2$ with $U(x')$ – as there are only a finite number t of moves he cannot run out of possible x'.

In this instance the countable models of T are particularly simple – indeed the theory T is $\aleph_0$-categorical, all countable models were isomorphic. In future more complex situations this will generally not be the case and indeed we find the study of the countable models of the almost sure theories T to be quite intriguing in its own right.

0.3 Some Comments on References

We have attempted to make this work, for the most part, self-contained, with definitions and theorems beginning with first principles. The subject matter itself has a rich literature. Here we discuss some of the most basic references and some excellent general references for the interested reader.

Our fundamental reference is two-fold, mathematics being affected by the social climate of the Cold War. The Zero-One Law (Theorem 0.1.2) was originally proven by Glebskii, Kogan, Liagonkii and Talanov [8] in 1969. It was proven independently by Fagin [7] in 1976. This author, like many on his side of the Iron Curtain, first heard of this result through Fagin's work. The two approaches to the result are quite different. For this author, it has been the insights of the Fagin approach that have led to the further results that make up this work.

The subject of Random Graphs had a clear beginning, with a fundamental paper [5] by the Hungarians giants Paul Erdős and Alfred Rényi in 1960. The work of Béla Bollobás [2] and the work of Svante Janson, Tomasz Łuczak, and Andrzej Ruciński [9] – both bearing the title *Random Graphs* – are fundamental references. This author would be remiss not to include his own joint work with Noga Alon [1] entitled *The Probabilistic Method.* This subject, developed by Paul Erdős, intertwines with Random Graphs and the work includes sections on Random Graphs and, most particularly, Zero-One Laws.

The central result of this work, Theorem 1.4.1, giving a Zero-One Law when $p = n^{-\alpha}$, α irrational, was first shown by Saharon Shelah and this author [15] in 1988. The approach via Ehrenfeucht games was given by this author [17] in 1991. Surveys of Zero-One Laws have been given by Kevin Compton [3], Peter Winkler [23] and this author [19].

One exception to our self-containment policy is Janson's Inequality, Theorem 5.0.4. This first appeared in [10], full treatments of the result, its generalizations and implications, appear in [9], [1] and elsewhere.

1. Preliminaries

First, what is a graph. We'll be looking at graphs on vertex set $V = \{1, \ldots, n\}$. For a combinatorialist, our graphs will be undirected, with neither loops nor multiple edges. For a logician, our graph is a set V together with a symmetric irreflexive binary relation. We express the relation – the notion that i, j are joined by an edge – by writing $i \sim j$. The terms "i, j are adjacent", "i, j are joined by an edge" and $i \sim j$ all have the same meaning.

1.1 What is the Random Graph $G(n, p)$?

Let's start with a thought experiment. We have n vertices labelled $1, \ldots, n$. Each unordered pair of vertices decides whether they are to be adjacent by flipping a coin. The coin is biased to come up heads with probability p and the edge is placed between them exactly when the coin comes up heads. This, on an intuitive level, is the random graph $G(n, p)$.

More formally we have a probability space whose elements are the $2^{\binom{n}{2}}$ labelled graphs on $V = \{1, \ldots, n\}$. The probabilities are determined by saying $\Pr[i \sim j] = p$ and that the events $i \sim j$ are mutually independent. Alternatively, if H is a graph with e edges then $\Pr\big[\{H\}\big] = p^e\,(1-p)^{\binom{n}{2}-e}$. However, we shall avoid this formulation as it disguises the critical role of the independence. We should point out that $G(n, p)$ is a finite probability space in which every singleton, and hence every set, does have a measure. Nonmeasurability plays no role here.

Let A be a property of graphs. Given a probability space $G(n, p)$ we naturally associate A with the event that the property A holds. We shall use the terms property and event interchangeably. Then $\Pr\big[G(n, p) \models A\big]$ is the probability of the event A in the probability space $G(n, p)$. More intuitively, suppose $G(n, p)$ is constructed by the coin flips as described above, then $\Pr\big[G(n, p) \models A\big]$ is the probability that that experiment will yield a graph with property A. Note that in the special case $p = \frac{1}{2}$ discussed in Section 0.1 all graphs have the same probability and so $\Pr\big[G(n, \frac{1}{2}) \models A\big]$ is precisely the proportion $\mu_n(A)$ of labelled graphs on n vertices satisfying property A.

1.1.1 The Erdős-Rényi Evolution

As p goes from zero to one the random graph $G(n,p)$ evolves from empty to full. For monotone properties A the value $\Pr\big[G(n,p) \models A\big]$, as a function of p, increases from zero to one. In their seminal work that began the field of Random Graphs, Paul Erdős and Alfred Rényi discovered that for many natural properties A there was a narrow range in which $\Pr\big[G(n,p) \models A\big]$ moved from near zero to near one. Critically, that range was generally not a constant p but moved as a natural function $p = p(n)$. They called that function $p(n)$ a threshold function for the property A.

As an instructive example, let A be the event "containing a triangle" first considered in Section 0.1. There are $\binom{n}{3}$ potential triangles and each has probability p^3 of being a triangle so the expected (average) number of triangles is $\binom{n}{3}p^3$. When $p = p(n) \ll \frac{1}{n}$ (e.g.: $p = n^{-1.01}$) the expected number of triangles is going to zero and so $\Pr[A]$ is going to zero. When $p = p(n) \gg \frac{1}{n}$ (e.g.: $p = n^{-0.99}$) the expected number of triangles is going to infinity. While this does not *a priori* give us information about $\Pr[A]$ it does turn out, as we shall see in Chapter 5, that $\Pr[A]$ goes to one. Erdős and Rényi called $p = \frac{1}{n}$ a threshold function for property A. More generally, they gave the following definition.

Definition 1.1 *A function $p(n)$ is a threshold function for A if*
- *If $p'(n) \ll p(n)$ then* $\Pr\big[G\big(n,p'(n)\big) \models A\big] \to 0$ *and*
- *If $p'(n) \gg p(n)$ then* $\Pr\big[G\big(n,p'(n)\big) \models A\big] \to 1$

Note that this definition ignores constant factors, so that we could equally well have said that $\frac{10}{n}$ is a threshold function for A.

It is useful, if not entirely accurate, to think of $G(n,p)$ as evolving from empty to full as $p = p(n)$ evolves through ever increasing functions of n. We state some results without proof to give the reader a feeling for this evolution. At $p = n^{-2}$ edges appear, by which we mean $p = n^{-2}$ is the threshold function for the property of being nonempty. At $p = n^{-3/2}$ edges with a common vertex appear. At $p = n^{-1-1/k}$ (with k arbitrary but fixed) all trees with $k+1$ vertices appear. At $p = \frac{1}{n}$ triangles appear, as do cycles of every fixed size k. Also, the graph becomes nonplanar. At $p = \frac{\ln n}{n}$ the graph becomes connected. At $p = n^{-2/3}$ complete subgraphs on four vertices appear and (with $k \geq 3$ fixed) at $p = n^{-2/(k-1)}$ complete subgraphs on k vertices (denoted K_k) appear. At $p = n^{-1/2}\ln^{1/2} n$ every pair of vertices gets a common neighbor. The study of these threshold functions has been, and continues to be, a central element in the study of Random Graphs.

In the above list the threshold functions seem to share common characteristics. They all seem to be powers of n, possibly times polylogarithmic terms. And, what turned out to be the central observation, *the powers of n in the threshold functions were always rational numbers.* That is, $n^{-\pi/7}$ never came up as a threshold function.

Hold on a moment. Let A be the property that the graph has at least $\binom{n}{2}n^{-\pi/7}$ edges. Its easy to show that $p = n^{-\pi/7}$ *is* the threshold function for that A. Still, this example does not seem natural. We would like to say that $n^{-\pi/7}$ never comes up as a threshold function for a natural property A. This shall be the fundamental result of this book.

1.1.2 The Appearance of Small Subgraphs

Erdős and Rényi studied carefully the appearance of an arbitrary but fixed graph H in the evolution of the random graph $G(n,p)$. Suppose H has v vertices and e edges. The number of potential copies of H is $\sim cn^v$ where c is a constant depending on the automorphism count. Each potential copy is a copy with probability p^e. Thus the expected number of copies of H is $\sim cn^v p^e$. When $p \ll n^{-v/e}$ this is $o(1)$ and almost surely H has not yet appeared. What if $p \gg n^{-v/e}$? The expected number of copies goes to infinity but this does not, a priori, mean that almost surely there will be a copy. Indeed, consider a "fish" consisting of a K_4 on vertices a, b, c, d and a tail vertex e adjacent only to d. The fish has $v = 5$ vertices and $e = 7$ edges. Will it appear near $p = n^{-5/7}$. Most definitely not – in order for the fish to appear the subgraph K_4 has to appear and it does not appear until $p \sim n^{-2/3}$ which is later!

Erdős and Rényi showed that, other then blockages from subgraphs, graphs H will appear when their expected number of copies becomes large. In general, among all the subgraphs H' of H (including H itself) compute the ratio e'/v' and take the maximal value. Then, they showed, the threshold function for the appearance of H is $n^{-v'/e'}$. We note that this is indeed a rational power of n.

Definition 1.2 *A graph H is* balanced *if the maximal ratio e'/v' of edges to vertices amongst all subgraphs is achieved for H itself and* strictly balanced *if that ratio was achieved only for H.*

The fish is not balanced as $6/4 > 7/5$. We can define a "house" consisting of a K_4 on vertices a, b, c, d plus two vertices e, f with a path $cefd$ forming the roof. The house is balanced but not strictly balanced as K_4 is a subgraph and $9/6 = 6/4$. The Erdős-Rényi theory of the appearance of general graphs nicely splits into problems for balanced and strictly balanced graphs. This theory provided strong motivation for much of the results presented here.

1.2 What is a First Order Theory?

The first order theory is a basic notion for logicians. Here we restrict ourselves to the first order theory of graphs. Basically, this is a language in which one can describe some, but certainly not all, properties of graphs.

The alphabet consists of

1. An infinite supply of variable symbols $x, y, z, \ldots$.
2. The relations $=$ (equality) and $\sim$ (adjacency). These can be used only between two variable symbols. E.g.: $x = y$ or $y \sim z$.
3. Universal $\forall$ and existential $\exists$ quantification. These can be used only on variable symbols. E.g.: $\forall_x, \exists_y$.
4. The usual Boolean connectives: $\vee, \wedge, \neg, \Rightarrow$.

What properties can we express in this language?

- There exists a triangle: $\exists_x \exists_y \exists_z [x \sim y \wedge y \sim z \wedge x \sim z]$
- There are no isolated vertices: $\forall_x \exists_y x \sim y$
- The radius is at most two: $\exists_x \forall_y [y = x \vee y \sim x \vee \exists_z [x \sim z \wedge z \sim y]]$

In the above expressions the variable symbols x, y, z are all what are called bound variables – they don't actually have values themselves but serve as place holders, much as the i in $\sum_{i=1}^{5} i^2$. There are other expressions, like $\exists_y x \sim y$ in which a variable (here x) is what is called a free variable. We can't say the truth of such an expression for a given graph G as it depends on what x is. The expressions with no free variables are called sentences, and these are the ones that correspond to properties of graphs. Still, the other expressions will prove important. The general term for an expression that may or may not have free variables is *predicate*. Predicates can be rigorously defined inductively: start with $x = y$ and $x \sim y$ (for any two variable symbols) and join predicates by Boolean connectives and by prefixing by universal or existential quantification. To prove a theorem about all sentences one often generalizes it to a theorem about all predicates and then proves that by induction on the length of the predicate: it suffices to show that the theorem holds for $x = y$, $x \sim y$ and that if it holds for P, Q then it holds for $\neg P$, $P \vee Q$ and (the most difficult and interesting case) $\exists_x P$.

The *quantifier depth* of a predicate is formally defined by induction. The atomic predicates $x = y$ and $x \sim y$ have quantifier depth zero. The quantifier depth of $\neg P$ is the quantifier depth of P. The quantifier depth of $P \wedge Q$, $P \vee Q$ and $P \Rightarrow Q$ are all the maximum of the quantifier depths of P and Q. Finally, and critically, the quantifier depth of $\exists_x P$ and $\forall_x P$ are both one more than the quantifier depth of P. Informally we think of quantifiers nested inside each other. For example, consider

$$(\exists_v v = v) \wedge \forall_x \Big[(\exists_y y \sim x) \wedge (\exists_z (\neg(z = x) \wedge \neg(z \sim x))) \Big] \tag{2}$$

Both $\exists_y$ and $\exists_z$ are nested inside $\forall_x$ but neither is nested inside the other. No quantifiers are nested inside $\exists_v$. The quantification depth is two. What graphs G have this property? The first part forces G to be nonempty. The main part says that all vertices have neighbors and nonneighbors. The G with this property form an equivalence class under $\equiv_2$ as defined in Section 2.2.

There are several important limitations of the first order language. First, the variable symbols x, y, z represent vertices – we cannot say that there exists a set X of vertices with a given property. (These come in in what are called second order languages. There one can say that a graph is three colorable by saying there exist sets Red, Blue, Green such that every vertex x is in one of them and if x, y are in Red (or Blue, or Green) then $\neg(x \sim y)$. This is a much stronger language. Outside of Chapter 11 we shall restrict ourselves to first order languages.) A second key limitation is that all sentences must be finite. Suppose we want to express connectivity. We want that $\forall_x \forall_y$ there is a path from x to y. For any fixed integer k we can say that all pairs of vertices are connected by a path of length k by writing

$$\forall_x \forall_y \exists_{z_1} \ldots \exists_{z_k} \left[x \sim z_1 \wedge z_1 \sim z_2 \wedge \ldots \wedge z_{k-1} \sim z_k \wedge z_k \sim y\right]$$

However, there is no single sentence that expresses connectivity. We cannot put ellipses into the sentence, each variable must be listed. [You might say: haven't we just done that. But for any particular value of k the expression above can be written out in full. This subtle distinction will come up often, as we will be exploring what sentences can be written in the first order language.] A third limitation is that the number n of vertices does not appear in the language. We can't say that n is even or that the graph contains a clique of size $\lfloor 2 \log_2 n \rfloor$, we can't even say the letter n.

We should note that in the above examples: three-colorability, connectedness, etc., we haven't really shown that these properties aren't expressible in the first order world, only that a natural attempt to express them fails because of limitations of the language. Proving that a property is not expressible by a first order sentence can be a tricky business and we'll have more to say about it in Section 2.3.

For typographical convenience we shall occasionally use $\bigvee, \bigwedge$ in place of $\vee, \wedge$ respectively. Conjunctions and disjunctions over designated index sets have their usual meaning, viz.

$$\bigwedge_{1 \le i < 4} x_i \neq x_{i+1} \text{ means } (x_1 \neq x_2) \wedge (x_2 \neq x_3) \wedge (x_3 \neq x_4)$$

1.3 Extension Statements and Rooted Graphs

A special role is played in the theory by a class of first order statements that we call extension statements. They include
- Every vertex lies in a triangle
- Every pair of vertices is connected by a path of length six
- Every three vertices have a common neighbor

and such "degenerate forms" as
- There exists a complete subgraph on 4 vertices.

Definition 1.3 *A* rooted graph *is a pair* (R, H) *where* $H = (V(H), E(H))$ *is a graph and* $R \subset V(H)$*,* $R \neq V(H)$*. The vertices of* R *are called roots.*

We allow R to be empty, indeed the rooted graphs $(\emptyset, H)$ play and important role. Let the vertices of H be labelled $a_1, \ldots, a_r; b_1, \ldots, b_v$ where the a_i are the roots and the b_j are the nonroots.

Definition 1.4 *The (R, H) extension statement, denoted* $\mathrm{Ext}(R, H)$, *is given by*

$$\forall_{x_1} \cdots \forall_{x_r} \left[\bigwedge_{i \neq i'} x_i \neq x'_i \Rightarrow \exists_{y_1} \cdots \exists_{y_v} \left[\bigwedge_{j \neq j'} y_j \neq y'_j \wedge \bigwedge_{i,j} x_i \neq y_j \wedge B \right] \right]$$

with the central part B defined as

$$B: \bigwedge_{\{a_i, b_j\} \in H} x_i \sim y_j \wedge \bigwedge_{\{b_j, b_k\} \in H} y_j \sim y_k$$

Informally: any r vertices can be extended to a copy of H with the r vertices in designated positions except

1. Edges between the roots do not count
2. The copy of H may have additional edges besides those of H

In the degenerate case $R = \emptyset$, $\mathrm{Ext}(R, H)$ becomes a purely existential statement, that there is a copy (by which we do not require an induced copy) of H.

Let G be any graph. Let $\mathbf{x} = (x_1, \ldots, x_r)$ be an r-tuple of distinct vertices of G and let $\mathbf{y} = (y_1, \ldots, y_v)$ be a v-tuple of vertices of G, distinct from each other and the x_i. We isolate the central part B of the statement $\mathrm{Ext}(R, H)$ as follows:

Definition 1.5 $\mathbf{y}$ *is an (R, H)*-extension of $\mathbf{x}$ *if $y_j \sim y_k$ whenever $\{b_j, b_k\} \in H$ and $x_i \sim y_j$ whenever $\{a_i, b_j\} \in H$.*

Thus $\mathrm{Ext}(R, H)$ holds for G if and only if every $\mathbf{x}$ has an (R, H)-extension $\mathbf{y}$. We note that $\mathbf{y}$ being an (R, H) extension of $\mathbf{x}$ depends on the labelling given to the vertices of H. For convenience, we imagine some labelling predetermined. The sentence $\mathrm{Ext}(R, H)$, however, does not depend on the labelling.

1.4 What is a Zero-One Law?

Let an edge probability function $p(n)$ be given. This gives a sequence of probability spaces $G(n, p(n))$.

Definition 1.6 *We say property A holds almost surely (relative to the function $p(n)$) if*

$$\lim_{n \to \infty} \Pr\left[G(n, p(n)) \models A\right] = 1$$

and almost never (relative to the function $p(n)$) if

$$\lim_{n\to\infty} \Pr\bigl[G\bigl(n, p(n)\bigr) \models A\bigr] = 0$$

or, equivalently, when $\neg A$ holds almost surely.

This generalizes the use of these terms in Definition 0.1, which applied to the special case $p = \frac{1}{2}$. But now the notion of almost always/never depends strongly on the choice of $p(n)$.

Definition 1.7 *We say $p(n)$* satisfies the Zero-One Law *if, relative to it, every first order sentence A holds either almost always or almost never.*

Theorem 0.1.2 then says that the constant function $p(n) = \frac{1}{2}$ satisfies the Zero-One Law. With Theorem 9.1.4 we shall rather easily generalize this result to all constant $p(n) = p \in (0,1)$ and somewhat beyond. This result still limits us to $p(n)$ that are, from the point of view of Random Graph Theory, rather restrictive. We can now state the central result of this work.

Theorem 1.4.1 (Our Main Theorem). *If $0 < \alpha < 1$ and α is irrational then $p(n) = n^{-\alpha}$ satisfies the Zero-One Law.*

The situation with A the property "contains a triangle" is instructive. The threshold function is $1/n$. When $p(n) \ll 1/n$ A holds almost never, when $p(n) \gg 1/n$ A holds almost always. When $p(n) = c/n$ we are "inside" the threshold interval and the probability is moving (with c) from zero to one. The precise result (see Chapter 5) is that for $p(n) = c/n$ $\Pr[A] \to 1 - e^{-c^3/6}$. The key, for our purposes, is that this is neither zero nor one. Thus $p(n) = c/n$ does not satisfy the Zero-One Law. Indeed, for $p(n)$ to satisfy the Zero-One Law we must have $p(n) \ll 1/n$ or $p(n) \gg 1/n$.

We are tempted to generalize. Take a monotone A with threshold function $p_A(n)$. There will be a threshold interval where $\Pr[A]$ goes from zero to one. If $p(n)$ is before that interval then A occurs almost never, if it is after then A occurs almost always. Our feeling is: the Zero-One Law holds when $p(n)$ is not in one of those threshold intervals, when $p(n)$ falls "between the cracks" of the threshold functions p_A. The Main Theorem 1.4.1 fits this view. If the threshold functions are basically rational powers of n then an irrational power would fall between them – their Dedekind Cut would provide a Zero-One Law. The threshold functions p_A are the "interesting" places in the evolution of the Random Graph and the Zero-One Law holds for the other functions. When told that $p(n) = n^{-\pi/7}$ satisfies the Zero-One Law some respond: What is so special about $n^{-\pi/7}$. The answer: Nothing! To show $p(n)$ satisfies the Zero-One Law is to show that $p(n)$ is a boring function.

A word of caution. Not all first order A are monotonic and not all first order A have threshold functions. So the notion of the Zero-One Law $p(n)$ being the antithesis of the threshold functions is not a rigorous one. Still, it provides a powerful intuitive understanding.

1.5 Almost Sure Theories and Complete Theories

Let $p(n)$ be any edge probability function.

Definition 1.8 *The almost sure theory T (relative to $p(n)$) is the set of all sentences A holding almost surely. We write T_α for the special case $p(n) = n^{-\alpha}$ when $0 < \alpha < 1$ is irrational.*

A theory is a set of sentences which are closed under logical inference in the first order language. We justify Definition 1.8 by the following basic result.

Theorem 1.5.1. *The almost sure theory T as given by Definition 1.8 is closed under logical inference. Furthermore, it is consistent.*

Proof: Since proof is finite any logical inference can only use (critically) a finite number of $A \in T$. Let $A_1, \ldots, A_s \in T$ and suppose B can be deduced from the A_i by first order logic. We have

$$\Pr\big[G(n,p(n)) \models \neg A_1 \vee \ldots \vee \neg A_s\big] \leq \sum_{i=1}^{s} \Pr\big[G(n,p(n)) \models \neg A_i\big]$$

The right side is a finite sum of terms, each approaching zero so it approaches zero and so the left side approaches zero. That is, $\neg A_1 \vee \ldots \vee \neg A_s$ holds almost never and so its negation $A_1 \wedge \ldots \wedge A_s$ holds almost always. Whenever it holds so does B, so B holds almost always. Further, T is consistent as the sentence "false" does not occur almost always (indeed, it never holds) and so is not in T.

Definition 1.9 *A theory T is called complete if for every sentence B either B or $\neg B$ is in the theory.*

We have a simple connection: $p(n)$ satisfies the Zero-One Law if and only if the theory T relative to $p(n)$ is complete.

The complete theories are of interest by themselves. From our Main Theorem 1.4.1 the theories T_α are complete. We shall explore natural axiomatizations of T_α in Section 7.1

1.6 Countable Models

Suppose $p(n)$ satisfies the Zero-One Law and let T be its almost sure theory. T is then a complete consistent theory. What kind of models does T have? [A model G of a theory T is simply a specific graph that satisfies all the properties $A \in T$.] We first note T can have no finite models. For any k there is a first order sentence with the meaning that the graph has at least k elements, namely

$$\exists_{x_1} \cdots \exists_{x_k} \bigwedge_{1 \leq i < j \leq k} x_i \neq x_k$$

This sentence holds almost surely since it holds with probability one for all $n \geq k$. Thus G would have to satisfy it and so have at least k elements. But this is true for all k and so G cannot be finite.

Does T have models? A deep theorem of Kurt Gödel, the Gödel Completeness Theorem, says yes, and more. It says that any consistent theory must have a countable or finite model. In our case, T must have a countable model. Such a countable model could be thought of, in the logical sense, as the limit of the probability spaces $G(n,p)$. Yet from a probabilistic point of view it is not clear how the countable model G would come from the $G(n,p)$. If, as is usually the case, $p(n) \to 0$ we do not see a way of directly defining a probability space on countable graphs that would reflect the asymptotics of $G(n,p)$.

A theory T is called $\aleph_0$-categorical (read: aleph nought categorical) if it has precisely one countable model up to isomorphism. Suppose T were not complete and $B, \neg B \notin T$. Adding B to T (and closing under first order inference) gives a theory T^+ while adding $\neg B$ gives a theory T^-. Both are consistent so, by Gödel, they have countable models G^+, G^-. These are both models of T but since they disagree on B's truth they cannot be isomorphic. Taking the contrapositive gives what is often denoted the Skolem-Lowenheim Theorem: If T has no finite models and is $\aleph_0$-categorical then T is complete. This is the result used in Section 0.1 to prove the Fagin-GKLT Theorem 0.1.2.

It will turn out that $\aleph_0$-categoricity is a rarity – the theories T_α do not have that property. Nonetheless, we will specify in Section 7.2 a specific countable model G_α of T_α with intriguing "minimal" properties.

Definition 1.10 *Two (infinite) graphs G_1, G_2 are called* elementarily equivalent *if they satisfy exactly the same first order properties.*

Naturally, if G_1, G_2 are isomorphic then they are elementarily equivalent. The converse, however, is very false. There are many examples (for example, Theorem 3.3.2) of nonisomorphic graphs which are elementarily equivalent. We can use the countable models of T to prove the completeness of T even in cases when T is not $\aleph_0$-categorical.

Theorem 1.6.1. *Let T be a theory with no finite models. Then T is complete if and only if all infinite graphs G satisfying T are elementarily equivalent.*

Proof: Suppose T is complete. For any first order A either $T \models A$ or $T \models \neg A$ and so either all models G satisfy A or all models G satisfy $\neg A$. Inversely, suppose T is not complete. There is a first order B with neither B nor $\neg B$ derivable from T. Then, as above, adding B and $\neg B$ give T^+ and T^- with models (by Gödel) G^+ and G^-. These are both models of T but as they disagree on B they are not elementarily equivalent.

We shall see in section 2.4 and elsewhere how we may sometimes use the Ehrenfeucht game to prove that any two models of T are elementarily equivalent.

2. The Ehrenfeucht Game

The Ehrenfeucht Game is a powerful tool for logicians but it can be an even more powerful tool for those with little knowledge of logic. The Game itself is defined without reference to Logic and the analysis of the game is combinatorial. Some basic bridging theorems allow one oftentimes to go from analysis of the Game to results about first order statements. Ehrenfeucht's original paper [6] is a true classic.

2.1 The Rules of the Game

The Ehrenfeucht Game is played by two players, called Spoiler and Duplicator. To further help distinguish them we make them male and female respectively. The Game has a certain number of rounds k which is known to both players. The Game has two graphs, which we shall call G_1, G_2 which are again known to both players. The two graphs will be on disjoint vertex sets. (Neither graph is "owned" by either player, we think of the two graphs on a table and the two players at either side.) These parameters determine the game which we shall call $\text{EHR}(G_1, G_2; k)$.

Each round has two parts, Spoiler's move followed by Duplicator's move. On the i-th move Spoiler selects a vertex on either graph (his choice) and marks it i. Then Duplicator must select a vertex on the other graph and also mark it i. A vertex may receive more than one mark.

Who wins? At the end of the game let $x_1, \ldots, x_k$ be the vertices of G_1 marked $1, \ldots, k$ respectively [regardless of who put the mark there] and let $y_1, \ldots, y_k$ be the corresponding vertices of G_2. For Duplicator to win she must assure that, for all $1 \leq i < j \leq k$, x_i, x_j are adjacent if and only if y_i, y_j are adjacent. As a more technical point, she must also assure that $x_i = x_j$ if and only if $y_i = y_j$. If Duplicator doesn't win then Spoiler wins.

That's it. We say $\text{EHR}(G_1, G_2; k)$ is a win for Duplicator if she wins with perfect play – that is, if there is a strategy for her so that regardless of what Spoiler does she wins. We say $\text{EHR}(G_1, G_2; k)$ is a win for Spoiler if, similarly, he wins with perfect play. The Ehrenfeucht Game is a finite perfect information game with no draws. Therefore it must be either a win for Duplicator or a win for Spoiler. Of course, like Hex and many challenging games, it might not be easy to give a winning strategy.

The case of marking the same vertex more than once can be completely disposed of. If Spoiler on his i-th move marks, say, $x_i = x_j$ then Duplicator simply sets $y_i = y_j$ and Spoiler has "wasted" a move. When Spoiler marks a new x_i Duplicator must also mark a new y_i or else she loses. Thus the game is equivalent if we require marks to be on new vertices. The one exception is when k is more than the number of vertices of G_1 (or G_2). Then some vertices must be marked more than once. These games can be completely described. If G_1, G_2 are isomorphic then Duplicator wins by duplicating. Otherwise Spoiler selects all of G_1 and Duplicator must select an isomorphic copy of G_1 in G_2. If such a copy exists but is not all of G_2 then Spoiler selects a new vertex of G_2 and Duplicator cannot follow suit in G_1. That is, when $k > \min(|G_1|, |G_2|)$ Duplicator wins if and only if G_1, G_2 are isomorphic.

Suppose on the i-th round Spoiler has selected $x_i \in G_1$. Duplicator knows $x_1, \ldots, x_{i-1}$, $y_1, \ldots, y_{i-1}$ and the latest x_i. She can see precisely which x_j, $j < i$ are adjacent to x_i. She must select a y_i with those same adjacencies to the previous y_j. If no such y_i exists then she loses. Even if such a y_i exists she may (and in important cases will) have to choose y_i carefully as different choices may allow different strategies for Spoiler later in the game. This does, however, give one very important case.

Theorem 2.1.1. *If G_1, G_2 both satisfy the Alice's Restaurant property and k is arbitrary then Duplicator wins* $\text{EHR}(G_1, G_2; k)$

Proof: Regardless of the previous history of the game Duplicator can always match x_i with a y_i (or y_i with x_i if Spoiler moves on G_2), and so she never gets stuck.

We've seen above that for fixed finite G_1, G_2 and k sufficiently large Duplicator wins if and only if the graphs are isomorphic. However, we shall basically be interested in cases where k is fixed and the graphs themselves become large. We think of, say, $k = 10$ and G_1, G_2 enormous graphs that are not isomorphic. Duplicator knows that there are only ten rounds and she tries only to "hang on" for that long.

Let's look at some situations that can be exploited by Spoiler. Suppose $k = 3$ and G_1 has a triangle but G_2 doesn't. Then Spoiler wins by selecting the three vertices of a triangle in G_1. More generally, if G_1 has an induced structure on k vertices that does not appear in G_2 then Spoiler wins the k move game by selecting that structure. The above example, while important, misses the essence of the Ehrenfeucht Game. *Spoiler's real strength comes from his ability to switch graphs.* The simplest situation is when G_1 has an isolated vertex v and G_2 does not and there are two moves. Spoiler first selects $x_1 = v$ and Duplicator must select some $y_1 \in G_2$. As G_2 had no isolated vertex y_1 has a neighbor and Spoiler selects it as y_2. Then y_2, y_1 are adjacent and Duplicator cannot find an x_2 adjacent to x_1 and so loses. Now suppose G_1 has radius two and G_2 has radius greater than two and there are three moves. Spoiler picks $x_1 \in G_1$ with all vertices of G_1 within distance two and Duplicator responds with some $y_1 \in G_2$. Spoiler switches graphs and picks $y_2 \in G_2$ at distance at least three from y_1 and Duplicator must pick $x_2 \in G_1$. She can't select x_2 adjacent to x_1 so x_1, x_2 must be at distance two. Now Spoiler switches again and selects x_3 as the common neighbor of x_1, x_2 and Duplicator is dead.

The case when G_1 and G_2 are both paths makes for an instructive example. Let P_n denote the path of length n and consider the game $\text{EHR}(P_n, P_m; k)$. The full analysis of this game is an interesting exercise but here we shall give suboptimal results that illustrate some general ideas. (The results on total orders, given in Section 2.6.2 are similar and, in some sense, purer and make for an interesting contrast.) We'll express the vertices as $\{1, \ldots, n\}$ and $\{1, \ldots, m\}$ respectively, with the understanding that these are disjoint sets. Suppose Spoiler plays an endpoint in either graph and there are at least two moves remaining. Then Duplicator must play an endpoint on the other graph, for if she picked some other vertex v Spoiler would select v's two neighbors on his next two moves and Duplicator would not be able to respond.

Theorem 2.1.2. *If $n \leq 2^k + 1$ and $n < m$ then Spoiler wins* $\text{EHR}(P_n, P_m; k+2)$.

Proof: We give a strategy for Spoiler in which he only moves in P_n. He first selects $x_1 = 1, x_2 = n$, the two endpoints, and by our comments above Duplicator must select y_1, y_2 as the endpoints of P_m so that $|y_2 - y_1| > |x_2 - x_1|$. We have $|x_2 - x_1| = n - 1 \leq 2^k$. Spoiler now employs a halving strategy, selecting x_3 as the midpoint (or either of the two midpoints if $x_2 - x_1$ is odd) of x_1, x_2. For any choice of y_3 either $|y_3 - y_2| > |x_3 - x_2|$ or $|y_2 - y_1| > |x_2 - x_1|$. Now there are some two marked points in P_n which are at most 2^{k-1} apart, so that the corresponding marked points in P_m are further apart then they are. Spoiler then plays the midpoint of these two points and continues. With s rounds remaining in the game Spoiler will have marked two points in P_n at most 2^s apart for which the corresponding marked points in P_m are further apart. By $s = 0$ these points are one apart, that is, they are adjacent. To win, Duplicator is required to preserve adjacency. Hence she has lost.

Theorem 2.1.3. *If $n, m \geq 2^{k+1} + 1$ then Duplicator wins* $\text{EHR}(P_n, P_m; k)$.

Proof: For convenience we'll allow Spoiler two free moves at the endpoints of P_n and have Duplicator respond with the endpoints of P_m. We'll thus assume $x_1 = 1, x_2 = n$ and $y_1 = 1, y_2 = m$ and that the game has $k + 2$ moves. Duplicator wants to be sure that when two marked vertices on one graph are close then their corresponding vertices are the same distance apart in the other graph. The key to the Duplicator strategy (here and elsewhere) is that the notion of close becomes tighter as the game approaches the end. Let s denote the number of rounds remaining in the game. When $s = 0$ (the end of the game) close means at distance one.

Suppose there are s moves remaining and the marked vertices are $x_i \in P_n$, $y_i \in P_m$, $1 \leq i \leq k + 2 - s$. Duplicator calls the positions equivalent if for every pair $1 \leq i < j \leq k+2-s$ either both x_i, x_j and y_i, y_j are more than 2^s apart or $x_j - x_i = y_j - y_i$. Her strategy will be to always move so that the positions remain equivalent. (That is, pairs that are more than 2^s apart are considered far apart and she doesn't care how far apart they are but pairs closer together have their corresponding pair in the same order and the same distance apart.) In the beginning the pairs are equivalent since x_1, x_2 and y_1, y_2 are both far apart. At the end (with $s = 0$) all pairs one apart have their corresponding pairs one apart. Adjacency has been preserved and Duplicator has won. The hard part – the essence of the proof – is the middle, showing that Duplicator can always move so that equivalent positions remain equivalent.

Suppose with s moves remaining the marked vertices $x_i \in P_n$, $y_i \in P_m$, $1 \leq i \leq k+2-s$ are equivalent and now Spoiler selects $x = x_{k+3-s} \in P_n$. (Selecting in P_m is similar.) We say (a notation that will appear again) Spoiler has moved *inside* if $|x - x_i| \leq 2^{s-1}$ for some $i \leq k+2-s$; otherwise we say Spoiler has moved *outside.*

Suppose Spoiler moved inside. Duplicator marks y with $y - y_i = x - x_i$. To see that the new positions are equivalent we must check where x or y are close to other vertices. Say y is close to y_j in that $|y - y_j| \leq 2^{s-1}$. But then

$$|y_i - y_j| \leq |y_i - y| + |y - y_j| \leq 2^{s-1} + 2^{s-1} = 2^s$$

Thus y_i, y_j were close with s moves remaining and so $y_i - y_j = x_i - x_j$. Thus $y - y_j = x - x_j$ as desired. When x is close to some other x_j the argument is identical. (Note how crucial it was in this part of the argument that the notion of closeness changed as s decreased.)

Suppose Spoiler moved outside. Duplicator marks any y that is outside, that is more than 2^{s-1} away from the previously marked vertices. We need only check that m is so large that there will necessarily be such a point and this we do by showing the number of inside points is less than m. When $s = k$ there are $2(2^{k-1}+1) = 2^k + 2 < m$ points within 2^{k-1} of the two endpoints. When $s = k-1$ there are at most $2(2^{k-1}+1) + (2^{k-1}+1)$ within 2^{k-2} of the three points (the nonendpoint having an interval roughly twice as large) and this is still less than m and for general s there are at most $2(2^{s-1}+1) + (k-s)(2^s+1)$ points within 2^{s-1} of the two endpoints and the $k-s$ nonendpoints. While the number of nonendpoints increases as s decreases (i.e., as the game progresses) the length of the intervals is halving each time and this number is always less than m.

For our purposes the essential point of the Theorem 2.1.3 was that if n, m were sufficiently large, dependent on k, then Duplicator wins $\text{EHR}(P_n, P_m; k)$. This is intuitively reasonable, with n, m very large the Spoiler does not have sufficient time to take advantage of the different distances between the endpoints. Still, Duplicator's strategy is surprisingly subtle, it changes as the game approaches its conclusion.

2.2 Equivalence Classes and Ehrenfeucht Value

In order to analyze the Ehrenfeucht Game it is helpful to consider a general intermediate position. Consider k-move Ehrenfeucht Game on G_1, G_2 and suppose $x_1, \ldots, x_s \in G_1$ and $y_1, \ldots, y_s \in G_2$ have already been marked. We may consider a game as starting in that position with $k - s$ rounds remaining. (If some x_i, x_j are adjacent but the corresponding y_i, y_j are not adjacent then poor Duplicator has lost before she began.) We write (with G_1, G_2 understood)

$$(x_1, \ldots, x_s) \equiv_k (y_1, \ldots, y_s) \tag{3}$$

if this game is a win for Duplicator. When $s = k$ the game has no moves. Duplicator wins if x_i, x_j are adjacent exactly when y_i, y_j are adjacent. Our original game may be considered the case $s = 0$ and we write

$$G_1 \equiv_k G_2 \tag{4}$$

when the game $\text{EHR}(G_1, G_2; k)$ is a win for Duplicator.

The relation $\equiv_k$ is reflexive (Duplicator duplicates Spoiler's move) and symmetric (the order of the graphs doesn't matter). We show transitivity by reverse induction on s, the case $s = k$ being clear from above. Assume the result for $s + 1$ and suppose we have three graphs with $(x_1, \ldots, x_s) \equiv_k (y_1, \ldots, y_s) \equiv_k (z_1, \ldots, z_s)$. In the game on G_1, G_3 suppose Spoiler plays some $x_{s+1} \in G_1$. Duplicator has a winning reply for the game on G_1, G_2 so there is a $y_{s+1} \in G_2$ with $(x_1, \ldots, x_{s+1}) \equiv_k (y_1, \ldots, y_{s+1})$. For the game on G_2, G_3 there is a winning reply $z_{s+1} \in G_3$ had Spoiler moved $y_{s+1} \in G_2$. That is, $(y_1, \ldots, y_{s+1}) \equiv_k (z_1, \ldots, z_{s+1})$. Spoiler responds to x_{s+1} by marking z_{s+1}. By induction on s we have $(x_1, \ldots, x_{s+1}) \equiv_k (z_1, \ldots, z_{s+1})$ so that Duplicator will win. Note that the final $s = 0$ case gives transitivity for $\equiv_k$ in the sense of 4.

Definition 2.1 *The* k-Ehrenfeucht value *of a graph G is the equivalence class it belongs to under $\equiv_k$. The k-Ehrenfeucht value of a graph G with marked $x_1, \ldots, x_s$ (where $s \leq k$) is the equivalence class it belongs to under $\equiv_k$. We let* $\text{EHRV}[k]$ *denote the set of Ehrenfeucht values for graphs and* $\text{EHRV}[k, s]$ *the set of Ehrenfeucht values for graphs with s marked vertices. We shall use the term Ehrenfeucht value of a graph when the particular value k is understood.*

Theorem 2.2.1. $\text{EHRV}[k]$ *and* $\text{EHRV}[k, s]$ *are finite sets. That is, there is a finite bound $F(k, s)$ for the number of equivalence classes under $\equiv_k$ for s-tuples.*

Proof. We prove this by reverse induction on s. When $s = k$ the equivalence classes are determined by the graph on the vertices $x_1, \ldots, x_k$. Thus we may use $F(k,k) = 2^{\binom{k}{2}}$ as an upper bound. Now suppose we have a bound $F(k,s+1)$ by induction. We claim we may take

$$F(k,s) = 2^{F(k,s+1)} \tag{5}$$

The equivalence class of $x_1,\ldots,x_s$ is determined by the set of possible $\equiv_k$ classes of $x_1,\ldots,x_s,x$ where x ranges over all vertices of G_1 other than $x_1,\ldots,x_s$. Why? Suppose $x_1,\ldots,x_s$ and $y_1,\ldots,y_s$ have the same set and Spoiler marks some $x \in G_1$. Duplicator can then play $y \in G_2$ with $(x_1,\ldots,x_s,x) \equiv_k (y_1,\ldots,y_s,y)$. Similarly, if Spoiler marks some $y \in G_2$ Duplicator has a response $x \in G_1$. Whatever Spoiler plays Duplicator has a winning response, hence Duplicator has a winning position. The number of equivalence classes of $x_1,\ldots,x_s$ is bounded by the number of subsets of the set of equivalence classes of $x_1,\ldots,x_{s+1}$ which is at most $2^{F(k,s+1)}$.

Our main application for Theorem 2.2.1 is when $s=0$: $\textsc{Ehrv}[k]$ is a finite set of size at most $F(k,0)$. In this case, as we shall soon see, finite does not mean small. We shall often use that $\textsc{Ehrv}[k]$ is finite – but for, say, $k=20$ it is far larger than the number of electrons in the universe. To understand the size of $\textsc{Ehrv}[k]$ we introduce a notation that is somewhat standard.

Definition 2.2 *The* tower *function, denoted $T(k)$, has domain the nonnegative integers and is given by initial value $T(0)=1$ and the recursion $T(k+1) = 2^{T(k)}$. The* logstar *function, denoted $\log^*(n)$, has domain the positive integers and is the least k with $n \le T(k)$.*

We may think of $T(k)$ as an exponential tower of twos of size k and $\log^*(n)$ as the number of times, beginning with n, that one needs to take the base 2 logarithm before reaching one or less. Now consider the upper bound for $F(k,0)$. The function $2^{\binom{k}{2}}$ is really the least important part – in the induction from $s=k$ to $s=0$ the bound is exponentiated at each step. For definiteness, we bound $k \le T(\log^* k)$ so that $\binom{k}{2} \le 2^k \le T(1+\log^* k)$ and thus $2^{\binom{k}{2}} \le T(2+\log^* k)$. Applying Equation 5 k times gives the upper bound for Theorem 2.2.2. (We are cheating slightly as the graphs with fewer than k vertices each form their own class and have not been counted. There are surely fewer than 2^{k^2} such graphs. This additional additive term is negligible compared to the tower function and may be absorbed into the upper bound below with a slight technical effort.)

Theorem 2.2.2. *For $k \ge 20$*

$$T(k-2) \le \left|\textsc{Ehrv}[k]\right| \le T(k+2+\log^* k)$$

The condition $k \ge 20$ is certainly not best possible and is inserted only to avoid technical difficulties. While these bounds may at first blush seem quite tight the skeptical reader will note that $T(i+1)$ and $T(i)$ are miles apart. For the lower bound we construct a family of graphs, denoted $G[X]$, which will be mutually k-inequivalent. The graphs are best described in levels.

1. Level 0: Three vertices a, b, c. We set $S_0 = \{a, b, c\}$.
2. Level 1: Seven vertices, each joined to a distinct nonempty subset of S_0. We set S_1 equal the set of these vertices.
3. Level i, $1 \leq i \leq k-4$. One vertex for each nonempty subset of S_{i-1}, joined to those vertices. We set S_i equal the set of vertices at level i.
4. A vertex d. For each nonempty $X \subseteq S_{k-4}$ in $G[X]$ d is adjacent to the vertices of X. (This is the only dependency on X.)
5. (Bells and whistles) Vertices $a^1, \ldots, a^4$ forming a pentagon with a; $b^1, \ldots, b^6$ forming a heptagon with b; $c^1, \ldots, c^8$ forming a 9-cycle with c; vertices $d^1, \ldots, d^{10}$ forming an 11-cycle with d.

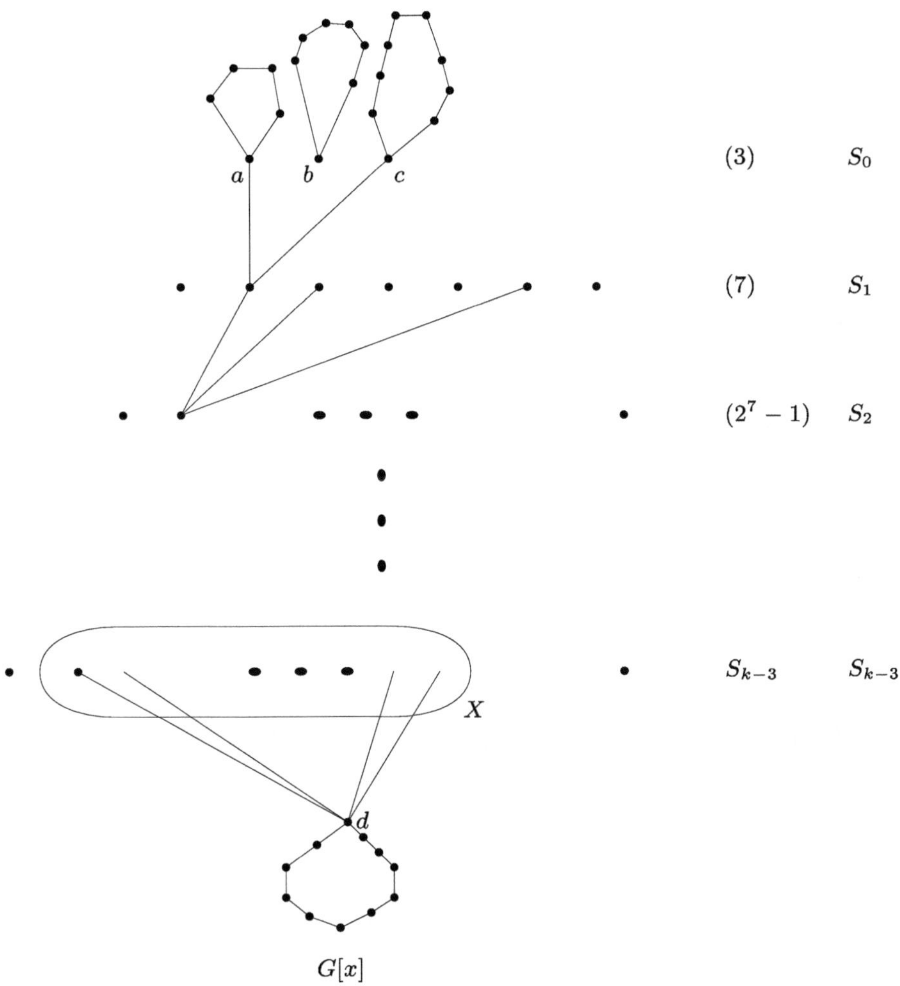

Fig. 2.1.

Let X_1, X_2 be distinct nonempty subsets of S_{k-4}. We shall show that $G[X_1] \not\equiv_k G[X_2]$ by giving an explicit strategy for Spoiler to win the k-move Ehrenfeucht game. The strategy is best described in stages.

1. (Initialization) Spoiler selects $a, b, c, d \in G[X_1]$. Here the bells and whistles force Duplicator to select $a, b, c, d \in G[X_2]$ in the same order. For suppose (other cases being similar) Duplicator responds to c with some other v. But c is the only vertex on a 9-cycle with a neighbor outside that cycle. Spoiler (as $k \geq 20$) would win by selecting that cycle and an outside neighbor on $G[X_1]$.
2. (Starting Down the Ladder) As $X_1 \neq X_2$ there is a vertex in one which is not in the other, say $v \in X_1$ and $v \notin X_2$. Spoiler, for his fifth move, selects $v \in G[X_1]$. As d, v are adjacent in $G[X_1]$ Duplicator must select $w \in G[X_2]$ adjacent to d. There are two cases:
   a) $w \notin S_{k-4}$. Then w is on the 11-cycle with d. There is a path of length $k-4$ from $v \in S_{k-4}$ to at least one of a, b, c (here we use that vertices on level i are adjacent to *nonempty* subsets of level $i-1$) but no such path from w. Spoiler completes the path with his remaining $k-5$ moves.
   b) $w \in S_{k-4}$. Then $v \neq w$ and we continue.
3. (Riding Down the Ladder) Formally this is an induction from $i = 1$ to $i = k - 4$. We show that Spoiler can force the $4 + i$-th moves to be two different vertices, both on the $k - 3 - i$-th level. The case $i = 1$ was just done. Now assume the induction hypothesis and let $v \in G[X_1]$, $w \in G[X_2]$ be the vertices chosen on the $k - 3 - i$-th round. Each is adjacent to a different subset of S_{k-4-i}, suppose, by symmetry, there is a $v' \in S_{k-4-i}$ adjacent to v but not to w. Spoiler now (for his $5 + i$-th move) selects $v' \in G[X_1]$. Duplicator must select $w' \in G[X_2]$ adjacent to w. There are two cases:
   a) $w' \notin S_{k-4-i}$. Then we must have $w \in S_{k-2-i}$ as adjacencies, except for the bells and whistles, are only between adjacent levels. There is a path of length $k-4-i$ from v' to at least one of a, b, c but no such path from w'. Spoiler completes the path with his remaining $k-5-i$ moves.
   b) $w' \in S_{k-4-i}$. Then $w' \neq v'$ and the induction hypothesis holds.
4. (Dead Duplicator) At $i = k - 4$, the $4 + i$-th, or final, move has distinct v, w from the first level selected. But a, b, c have already been selected in the same order in both graphs and v, w differ in their adjacencies to at least one of the a, b, c. Thus Duplicator has lost.

It remains to calculate the number of these graphs $G[X]$ as a function of k. Define a sequence $u_0, \ldots, u_{k-3}$ by initial value $u_0 = 3$ and recursion $u_{i+1} = 2^{u_i} - 1$. There are u_i vertices at level i for $0 \leq i \leq k - 4$ and hence u_{k-3} different nonempty $X \subseteq S_{k-4}$. The -1 term is only a minor irritant. We claim $u_i \geq T(i+1) + 1$ for all $i \geq 0$. This holds for $i = 0$ (which was why three vertices were placed at level 0) and if it holds for i then

$$u_{i+1} \geq 2^{T(i+1)+1} - 1 = 2T(i+2) - 1 \geq T(i+2) + 1$$

with room to spare. We neglect the +1 addend as we have neglected other factors in our favor to give the clean form of Theorem 2.2.2.

Let us turn now to the simpler case of $k = 2$. Pairs x_1, x_2 may be adjacent or not so that $|\text{EHRV}[2,2)| = 2$. Thus $|\text{EHRV}[2,1]| \leq 2^2 = 4$ but there are actually three classes. A vertex x_1 can be either isolated (all x_1, x_2 nonadjacent), focal (all x_1, x_2 adjacent) or mixed. Thus $|\text{EHRV}[2]| \leq 2^3 = 8$ but actually there are five kinds of graphs:

1. All vertices focal. (The complete graph.)
2. All vertices isolated. (The empty graph.)
3. All vertices focal or mixed, with some of each.
4. All vertices isolated or mixed, with some of each.
5. All vertices mixed.

Duplicator wins $\text{EHR}(G_1, G_2, k)$ if and only if G_1, G_2 are in the same category. To be a purist, the graphs on one and zero vertices (*not* a pointless concept!) are in their own categories.

Even for $k = 3$ a complete description of the equivalence classes is quite elaborate. Theorem 8.2.1 (basically the Trakhtenbrot-Vaught Theorem) indicates the logical limits of such a complete description, at least for those classes with finite models. One particular equivalence class deserves special attention.

Definition 2.3 *We say G has the k-Alice's Restaurant property if it has more than k vertices and if for all $x_1, \ldots, x_k \in G$ and all $\epsilon_1, \ldots, \epsilon_k \in \{0,1\}$ there exists z distinct from the x's with z adjacent to x_i if and only if $\epsilon_i = 1$. That is, for every k-set of vertices there exist witnesses z with every possible adjacency pattern to the k-set. Equivalently, G satisfies $A_{r,s}$, as defined in Section 0.1, for all nonnegative r, s with $r + s \leq k$.*

Theorem 2.2.3. *The family of graphs with the k-Alice's Restaurant property form a $\equiv_{k+1}$ class.*

Proof: If G_1, G_2 both have this property Duplicator wins as she can always find a vertex with matching adjacencies. Inversely, suppose G_2 has the property but in G_1 there are $x_1, \ldots, x_k$ and an adjacency pattern with no witness z. Spoiler selects $x_1, \ldots, x_k$ and Duplicator picks some $y_1, \ldots, y_k \in G_2$. Now Spoiler picks $y_{k+1} \in G_2$ having that adjacency pattern to the y's (such y_{k+1} exist by the k-Alice's Restaurant property) and Duplicator is dead.

2.3 Connection to First Order Sentences

Combinatorialists like games. Logicians like truth. Fortunately, there is a connection. A first order sentence, such as $\forall_x \exists_y x \sim y$ is both a formal sequence of symbols and a property that a graph may or may not have. When a graph G does have that property we say it has truth value true for that sentence, otherwise it has truth value false. The following core theorem provides a bridge from truth values to games.

Theorem 2.3.1. *For all $k \geq 1$:*

1. *$G_1 \equiv_k G_2$ if and only if G_1, G_2 have the same truth value for all first order sentences of quantifier depth k.*
2. *To each k-Ehrenfeucht value there is a first order sentence A of quantifier depth k such that the graphs G with that value are precisely those G for which A is true.*

For example, the 2-Ehrenfeucht value "all vertices mixed" discussed in Section 2.2 is just those graphs G satisfying (2). To prove Theorem 2.3.1 we actually show the following stronger result.

Theorem 2.3.2. *For all $k \geq 1$ and all $0 \leq s \leq k$:*

1. *$(G_1, x_1, \ldots, x_s) \equiv_k (G_2, y_1, \ldots, y_s)$ if and only if G_1, G_2 have the same truth value for all first order predicates of quantifier depth $k - s$ with s free variables, when the free variables are given the values $x_1, \ldots, x_s$ and $y_1, \ldots, y_s$ respectively.*
2. *To each k-Ehrenfeucht value on graphs with s designated vertices there is a first order predicate A of quantifier depth $k-s$ and s free variables such that the $(G, x_1, \ldots, x_s)$ with that value are precisely those G for which A is true when the free variables are given the designated vertices as values.*

Proof. When $s = k$ this is easy. Two graphs with k designated vertices are $\equiv_k$ if the induced subgraphs on their designated vertices are the same (not just isomorphic) and so the predicate A would list the adjacencies and nonadjacencies amongst the x_i. When $G_1 \equiv_k G_2$ all predicates with quantifier depth $k - s = 0$ are Boolean combinations of the atomic $x_i \sim x_j$ and $x_i = x_j$ and so would have the same truth value. When they are not $\equiv_k$ they differ on some adjacency and hence some atomic $x_i \sim x_j$ has different truth value on the two graphs.

Now assume the result, by induction, for $s + 1$ and consider the second part. Let $\alpha \in \textsc{Ehrv}[k, s]$. By induction every value $\beta \in \textsc{Ehrv}[k, s + 1]$ corresponds to a predicate $A_\beta(x_1, \ldots, x_s, x)$. Given α the truth value of $\exists_x A_\beta(x_1, \ldots, x_s, x)$ is determined. Let $\textsc{Yes}[\alpha]$ be those β for which the truth value is true, $\textsc{No}[\alpha]$ those β for which the truth value is false. Then we set $A_\alpha(x_1, \ldots, x_s)$ to:

$$\left[\bigwedge_{\beta \in \textsc{Yes}[\alpha]} \exists_x A_\beta(x_1, \ldots, x_s, x)\right] \wedge \left[\bigwedge_{\beta \in \textsc{No}[\alpha]} \neg\exists_x A_\beta(x_1, \ldots, x_s, x)\right]$$

When G with designated $x_1, \ldots, x_s$ satisfies A_α the set of Ehrenfeucht values with additional designated x is determines as $\textsc{Yes}[\alpha]$ and so G must have Ehrenfeucht value α.

Now for the first part. The A_α given above have, by induction, quantifier depth $k - s$ with s free variables. Suppose G_1, G_2 (with designated vertices) have the same truth value on each A_α. As $\bigvee_\alpha A_\alpha$ is a tautology they must both satisfy some A_α, hence both satisfy the same A_α, hence both have the same Ehrenfeucht value. Conversely, suppose G_1, G_2 (with designated vertices) are equivalent. Let P be any predicate of quantifier depth $k-s$ with s free variables. We may express P is a Boolean combination of predicates of the form $\exists_x Q$ where Q has quantifier depth $k-s-1$ with $s+1$ free variables. By induction the truth value of Q is determined by the Ehrenfeucht value of $(G, x_1, \ldots, x_s, x)$. Thus the truth value of $\exists_x Q$ is determined by α and hence the truth value of P is determined by α.

For those who love analyzing games but have little formal experience in logic Theorem 2.3.1 is the *only* result one really needs to prove results about First Order Sentences. We give some important applications that do not involve probability.

Definition 2.4 *A property P of graphs is called first order expressible if it can be written as a first order sentence A – i.e., so that A holds exactly when P holds.*

It is usually easy to show P is expressible by writing it in the first order language. To prove P is not expressible we turn Theorem 2.3.1 around.

Theorem 2.3.3. *Suppose that for all k there exists a pair of graphs G_1, G_2 with G_1 having property P and G_2 not having property P such that Duplicator wins* $\textsc{Ehr}(G_1, G_2; k)$. *Then P is not first order expressible.*

Proof: If P were first order expressible by some first order A that A would have to have some quantifier depth k. With G_1, G_2 disagreeing about A, Theorem 2.3.1 gives $G_1 \not\equiv_k G_2$. Thus, Spoiler wins $\textsc{Ehr}(G_1, G_2; k)$.

2.4 Inside-Outside Strategies

We give here several applications of Theorem 2.3.3 to show that various properties are not first order expressible. In all cases we need find a winning strategy for Duplicator in certain Ehrenfeucht games. These strategies will have many common features, we call them Inside-Outside Strategies. Theorem 2.1.3 provided a first example of this type of argument.

Theorem 2.4.1. *Connectivity is not first order expressible.*

Proof: Given k let G_1 consist of a cycle of order n and G_2 consist of two disjoint cycles of order n where we select $n \geq 2^{k-1}$. Our strategy for Duplicator mirrors that of Theorem 2.1.3. With s moves remaining in the game she calls two points close if they are at distance at most 2^s. She requires that if any two marked points are close on one graph then the corresponding points must be the same distance apart and in the same orientation in the other graph. More formally, we may label the vertices of each cycle by Z_n. Duplicator then requires that if moves $y_i, y_j \in G_2$ lie on the same cycle and $y_j = y_i + d$ with $0 < d < 2^s$ then the corresponding moves $x_i, x_j \in G_1$ have $x_j = x_i + d$. She also requires that if moves $x_i, x_j \in G_1$ have $x_j = x_i + d$. then the corresponding moves $y_i, y_j \in G_2$ lie on the same cycle and $y_j = y_i + d$. A new move $x \in G_1$ (moving in G_2 is similar) by Spoiler is called inside if it is within 2^{s-1} of a previous x_i and otherwise outside. If inside we have $x = x_i + d$ and Duplicator marks $y = y_i + d$. As in Theorem 2.1.3 this preserves the condition. If Spoiler moves outside Duplicator makes any move that is outside. With $k - s$ marked vertices there are at most $(k-s)[2^s - 1]$ vertices inside. As long as $n > (k-s)[2^s - 1]$ for all $1 \leq s \leq k-1$ (it happens $s = k-1$ maximizes this condition) there is always an outside point. Hence Duplicator can hold this condition to the end and at the adjacency is the same in both graphs so she has won.

We can picture the above argument with Spoiler beginning with a selection of one vertex in each of the cycles of G_2, trying to take advantage of G_2's nonconnectivity. But Duplicator counters by selecting two vertices of G_1 that are so far apart that Spoiler can't "prove" they have a path between them in the remaining time.

Theorem 2.4.2. 2-*Colorability is not first order expressible.*

Proof: The cycle C_n of size n is 2-colorable precisely when n is even. For any k let G_1,G_2 be cycles of size n,$n+1$ with $n \geq 2^{k-1}$. By the argument above Duplicator wins $\text{EHR}(G_1, G_2; k)$.

Theorem 3.3.3 gives a somewhat more elaborate nonexpressibility result.

We conclude this section with two somewhat more general situations in which Duplicator has a winning Inside-Outside strategy. Two further situations will be given by Theorems 2.6.6 and 2.6.7. We begin with notation.

Definition 2.5 *Let x, x' be distinct vertices of G. The* distance *from x to x', denoted $\rho(x, x')$ is the minimal t such that there exists a sequence $x = x_0, x_1, \ldots, x_t = x'$ with x_i, x_{i+1} adjacent for $0 \leq i < t$. Equivalently, $\rho(x, x')$ is the length of the shortest path between the vertices. When no such sequence exists we say $\rho(x, x') = \infty$. When $x = x'$ we define $\rho(x, x) = 0$.*

Definition 2.6 *Let $x \in G$, $d \geq 0$. The* d-neighborhood *of x is the set of x' with $\rho(x, x') \leq d$.*

Definition 2.7 *Let $x_1, \ldots, x_u \in G$, $d \geq 0$. The* d-picture *of $x_1, \ldots, x_u$ is the restriction of G to the union of the d-neighborhoods of the x_i.*

Definition 2.8 *Let $x_1, \ldots, x_u \in G_1$, $y_1, \ldots, y_u \in G_2$, $d \geq 0$. Their d-pictures are called* the same *if there is a graph isomorphism Ψ between the d-pictures that sends x_i to y_i for $1 \leq i \leq u$.*

Theorem 2.4.3. *Set $d = \frac{3^k - 1}{2}$. Suppose*

1. *For each $y \in G_2$ and each $x_1, \ldots, x_{k-1} \in G_1$ there is an $x \in G_1$ with x, y having the same d-neighborhood and such that $\rho(x, x_i) > 2d + 1$ for all $1 \leq i \leq k - 1$.*
2. *For each $x \in G_1$ and each $y_1, \ldots, y_{k-1} \in G_2$ there is an $y \in G_2$ with y, x having the same d-neighborhood and such that $\rho(y, y_i) > 2d + 1$ for all $1 \leq i \leq k - 1$.*

Then Duplicator wins $\text{EHR}(G_1, G_2; k)$.

Proof: Set $d_0 = 0$ and define inductively $d_s = 3d_{s-1} + 1$ so that $d = d_k$ above. Suppose $x_1, \ldots, x_{k-s} \in G_1$ and $y_1, \ldots, y_{k-s} \in G_2$ have been played. Duplicator's strategy is to have their d_s-pictures the same. If she succeeds in holding to this strategy then at $s = 0$, the end of the Game, $d_s = 0$ and the induced graphs on the union of the 0-neighborhoods, i.e., the marked vertices themselves, are isomorphic and she has won. She can certainly hold to this strategy for the first move since G_1, G_2 have the same d-neighborhoods. Suppose inductively that she can hold to this strategy until there are s moves remaining. For convenience set $D = d_{s-1}$ so $d_s = 3D + 1$ and there is an isomorphism between the induced graphs on the unions of the $3D + 1$-neighborhoods of the marked vertices. Now, by symmetry, say Spoiler plays $x = x_{k-s+1} \in G_1$.

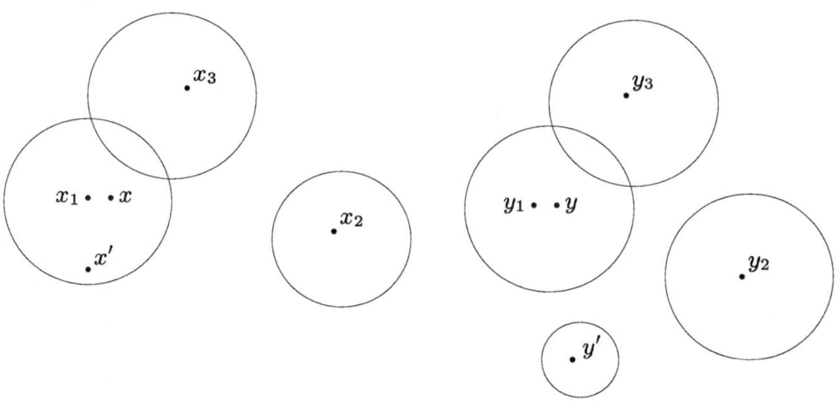

Fig. 2.2. When Spoiler plays x near x_1 (inside) Duplicator uses the isomorphism to play y. The trickiest aspect is when Spoiler plays x' more than $2D + 1$ from x. Duplicator regards this move as outside and finds y' possibly far from y. In the new picture (with smaller circles) x, x' are in separate components

We call x inside if it is at distance at most $2D+1$ from some previously marked vertex x_i, otherwise it is outside.

Suppose x is inside. Then the D-neighborhood of x is contained in the $3D+1$-neighborhood of x_i. Duplicator looks at the isomorphism ψ between the unions of the $3D+1$-neighborhoods and plays the corresponding $y = \psi(x) \in G_2$.

Suppose x is outside. In the new picture, the union of the D-neighborhoods of the marked vertices, we claim that the neighborhood of x will be a distinct component. For if x', x_i' in the D-neighborhoods of x, x_i were adjacent (or equal) then x, x_i would have been within distance $2D+1$. Now Duplicator moves to G_2 and marks a vertex y having the same D-neighborhood as x and at distance at least $2D+1$ for all the previously marked z_i.

Sometimes a graph G will have a set of "special" vertices S with unusual neighborhoods. (In application S will consist of the small cycles of G.) Our next result essentially says that if two graphs have appropriately identical such sets and otherwise satisfy the assumptions of Theorem 2.4.3 then they satisfy its conclusion as well.

Theorem 2.4.4. *Set* $d = \frac{3^k-1}{2}$. *Suppose* G_1, G_2 *have vertex subsets* $S_1 \subset G_1, S_2 \subset G_2$ *with the following properties:*

1. *The restrictions of* G_1 *to* S_1 *and* G_2 *to* S_2 *are isomorphic and this isomorphism can be extended to one between the d-neighborhoods of* S_1 *and* S_2.
2. *Let* $d' \le d$. *Let* $y \in G_2$ *with* $\rho(y, s_2) > 2d'+1$ *for all* $s_2 \in S_2$. *Let* $x_1, \ldots, x_{k-1} \in G_1$. *Then there exists* $x \in G_1$ *with* x, y *having the same* d'*-neighborhood and such that* $\rho(x, x_i) > 2d'+1$ *for* $1 \le i \le k-1$ *and* $\rho(x, s_1) > 2d'+1$ *for all* $s_1 \in S_1$.
3. *Let* $d' \le d$. *Let* $x \in G_1$ *with* $\rho(x, s_1) > 2d'+1$ *for all* $s_1 \in S_1$. *Let* $y_1, \ldots, y_{k-1} \in G_2$. *Then there exists* $y \in G_1$ *with* y, x *having the same* d'*-neighborhood and such that* $\rho(y, y_i) > 2d'+1$ *for* $1 \le i \le k-1$ *and* $\rho(y, s_2) > 2d'+1$ *for all* $s_2 \in S_2$.

Then Duplicator wins $\text{EHR}(G_1, G_2; k)$.

Duplicator imagines all $s_1 \in S_1$ and $s_2 \in S_2$ to be marked, the isomorphism giving the correspondence between the markings. There are k moves remaining. Duplicator then plays to assure that with i rounds remaining the d_i-neighborhoods of the marked vertices are the same. At $i = k$ this is true by assumption. Suppose with i moves remaining Spoiler selects (by symmetry) $x \in G_1$. Set $d_i = 3D+1$. Call x inside if it is at distance at most $2D+1$ from a previously marked vertex (including S_1), otherwise x is outside. If x is inside Duplicator follows the strategy of Theorem 2.4.3. If x is outside the third assumption allows Duplicator to again follow the strategy of Theorem 2.4.3. At $i = 0$ the game is over and Duplicator has won.

2.5 The Bridge to Zero-One Laws

Suppose the for every n we are given a probability space of graphs on the labelled vertices $1, \ldots, n$, which we call G_n. (In our work G_n will be the random graph $G(n, p(n))$.) We say that the sequence of spaces G_n satisfies the Zero-One Law if for every first order sentence A.

$$\lim_{n \to \infty} \Pr\left[G_n \models A\right] = 0 \text{ or } 1 \tag{6}$$

When $G_n \sim G\bigl(n, p(n)\bigr)$, the only case we shall consider in this work, the above notion is simply Definition 1.7.

Theorem 2.5.1 (Bridge Theorem). *The sequence G_n satisfies the Zero-One Law if and only if for every positive integer k*

$$\lim_{m,n \to \infty} \Pr\left[\textit{Duplicator wins } \text{EHR}(G_n, G_m; k)\right] = 1 \tag{7}$$

What does this mean? For a given n, m let G_n, G_m be independently chosen from their probability space and be (even when $n = m$) on disjoint vertex sets. Spoiler and Duplicator examine the G_n, G_m and, being infinitely intelligent, see who will win the k-move Ehrenfeucht game. (Very specifically, there is no sense of the players playing randomly – the players play perfectly, the board is random.) Backing up, there is a probability that G_n, G_m will be such that Duplicator claims victory. Backing up, it might be that the limit of this probability as $m, n \to \infty$ is one – i.e., for all positive ϵ there is an N so that for $m, n > N$ Duplicator will claim victory at least $1 - \epsilon$ of the time. Backing up, this might hold for any fixed length of the game k. (Of course, the N above might change with k.) That is the condition.

Assume 6 did not hold for some specific first order A. There would be a positive ϵ so that for all N there would be an $m > N$ with $\Pr\left[G_m \models A\right] > \epsilon$ (otherwise $\lim_{n\to\infty} \Pr\left[G_n \models A\right] = 0$) and an $n > N$ with $\Pr\left[G_n \models A\right] < 1 - \epsilon$ (otherwise $\lim_{n\to\infty} \Pr\left[G_n \models A\right] = 1$). Let k be such (using our fundamental Theorem 2.3.1) that Spoiler wins $\text{EHR}(H_1, H_2; k)$ whenever H_1 satisfies A and H_2 does not. As G_m, G_n are independent chosen the probability that G_m satisfies A and G_n does not is at least ϵ^2. So Spoiler wins $\text{EHR}(G_n, G_m; k)$ with probability at least ϵ^2. Thus 7 would fail.

Assume 6 does hold for every first order A. Let k be arbitrary but fixed. For each Ehrenfeucht value $\alpha \in \text{EHRV}[k]$ let A_α be that first order sentence that holds exactly for those G with value α. The A_α partition the probability space. Set $\epsilon_\alpha = \lim_{n\to\infty} \Pr\left[G_n \models A_\alpha\right]$ so that ϵ_α is always zero or one by our assumption. For $\alpha \neq \beta$ the events A_α, A_β are disjoint so we can't have both $\epsilon_\alpha = \epsilon_\beta = 1$. Further

$$1 = \sum_{\alpha \in \text{EHRV}[k]} \Pr\left[G_n \models A_\alpha\right]$$

Taking limits [making crucial use of the *finiteness* of EHRV$[k]$] gives $1 = \sum_\alpha \epsilon_\alpha$. Hence there is a unique α with $\epsilon_\alpha = 1$. Let $\epsilon > 0$. There exists n_0 such that for $n > n_0$ G_n has Ehrenfeucht value α with probability at least $1 - \epsilon$. For $n, m > n_0$ independently chosen G_n, G_m both have Ehrenfeucht value α with probability at least $(1-\epsilon)^2$. When this occurs Duplicator wins EHR$[G_n, G_m; k]$. As ϵ can be made arbitrarily small Equation 7 follows.

We generally will apply the Bridge Theorem 2.5.1 in the positive (if) sense. We shall for all k give a strategy for Duplicator that almost surely wins EHR$[G_n, G_m; k]$ and this will imply the Zero-One Law.

We illustrate this with an alternate proof of the fundamental Fagin-GKLT Theorem 0.1.2. G_n will have distribution $G(n, \frac{1}{2})$. Let k be arbitrary but fixed. Duplicator's strategy [for this particular example] will be quite simple – when a vertex x is marked by Spoiler she checks the adjacencies of x to the previously marked x_i and then goes to the other graph and marks a z with the same adjacencies to the corresponding z_i. This strategy clearly wins *if* it can be held to. We thus need only show that almost surely this strategy succeeds, that it can't be foiled by a wily Spoiler. We note that almost surely G_n has the $(k-1)$-Alice's Restaurant property given by Definition 2.3. When G_n, G_m both have the $(k-1)$-Alice's Restaurant property then Duplicator will win with the above strategy. Thus Duplicator almost surely wins EHR$[G_n, G_m; k]$, thus the Fagin-GKLT Zero-One Law.

We close this section with an important illustration of why this strategy does not work for the random graph $G(n, n^{-\alpha})$. We'll take $\alpha = \pi/7$. [For consistency of exposition we shall often use this value of α which is an irrational number with no particular properties.] Consider the game EHR$[G_n, G_m; 3]$ and suppose Duplicator announces she will be playing with no foresight but simply selecting a vertex with the desired adjacencies to previously marked vertices. Wily Spoiler notes [but does not mark] a vertex x in G_n and marks two of its neighbors as x_1, x_2. Duplicator marks some z_1, z_2 on G_2 but only keeping the adjacency or nonadjacency of the x's. Then Spoiler springs his trap – marking x as x_3. In G_m most pairs of vertices do not have a common vertex [as we shall see, and much more, in Section 5.1] so that unless Duplicator was lucky she has lost. Does this mean that the Zero-One Law fails? Not at all! We've merely shown that this particular strategy of Duplicator's does not win almost surely. Indeed, our proof of the Zero-One Law for this case will come through giving a more complicated strategy for Duplicator that does win almost surely.

2.6 Other Structures

2.6.1 General First Order Structures

While our emphasis through most of this work will be on graphs the Ehrenfeucht Game can be defined in a much more general setting. Let us first describe a general first order theory. The alphabet consists of

1. An infinite supply of variable symbols $x, y, z, \ldots$.
2. Universal $\forall$ and existential $\exists$ quantification. These can be used only on variable symbols. E.g.: $\forall_x, \exists_y$.
3. The usual Boolean connectives: $\vee, \wedge, \neg, \Rightarrow$.
4. A variety of relation symbols, which can be of varying arity. We shall always include equality as one of the symbols though this is not standard notation.
5. A variety of function symbols, which can be of varying arity.
6. A variety of constant symbols.

For graphs the relation symbols are $=$ (equality) and $\sim$ (adjacency). For ordered graphs as defined in Section 11.1 we also have the binary relation $<$ (less than) giving the ordering. For strings as defined in Chapter 10 we have equality, less than, and a unary predicate $U(x)$, indicating whether or not the bit in position x is a one. Total orders, as defined in Section 2.6.2 have only equality and less than. Most of the systems studied in this work have neither function symbols nor constant symbols. A standard example of a system with both these items is fields, with equality, binary function symbols $+$ and $\times$, and constant symbols $0, 1$.

Let M_1, M_2 be two models of a language and k a positive integer. Then we may define the Ehrenfeucht Game $\text{EHR}[M_1, M_2; k]$ between Spoiler and Duplicator. There are k rounds. Each round has two parts. On the i-th round Spoiler first selects a vertex in either model (his choice) and marks it i. Then Duplicator must select a vertex in the other model and mark it i. Formally, a vertex may receive more than one mark. Who wins? At the end of the game let $x_1, \ldots, x_k$ be the vertices of M_1 marked $1, \ldots, k$ respectively [regardless of who put the mark there] and let $y_1, \ldots, y_k$ be the corresponding vertices of M_2. Let R be a relational symbol of the language of arity s. (In all of our examples $s = 1$ or $s = 2$.) For Duplicator to win she must assure, for every such R, that for all $i_1, \ldots, i_s \in \{1, \ldots, k\}$ (allowing repetition) that $R(x_{i_1}, \ldots, x_{i_s})$ holds in M_1 if and only if $R(y_{i_1}, \ldots, y_{i_s})$ holds in M_2. Let F be a function symbol of the language with $s - 1$ variables. For Duplicator to win she must assure, for every such F, that for all $i_1, \ldots, i_s \in \{1, \ldots, k\}$ (allowing repetition) that $F(x_{i_1}, \ldots, x_{i_{s-1}}) = x_{i_s}$ holds in M_1 if and only if $F(y_{i_1}, \ldots, y_{i_{s-1}}) = y_{i_s}$ holds in M_2.

Constant symbols add a further restriction. Let R be a relational symbol of arity s. In M_1 plug in for each variable symbol either a marked vertex or a vertex representing a constant symbol. In M_2 plug in the corresponding marked vertices or vertices representing the same constant symbol. The truth value of R must then be the in both models. Functional symbols F add a similar requirement. These will not concern us in this work.

We write $M_1 \equiv_k M_2$ if $\text{EHR}(M_1, M_2; k)$ is a win for Duplicator. As in Section 2.2, $\equiv_k$ is an equivalence relation on the models. As in Definition 2.1 we may define the k-Ehrenfeucht value of a model M and the set $\text{EHRV}[k]$ of those values. As in Theorem 2.2.1, $\text{EHRV}[k]$ is a finite set. Finally the connection to First Order Sentences given by Theorem 2.3.1 generalizes:

Theorem 2.6.1. *For all $k \geq 1$:*

1. *$M_1 \equiv_k M_2$ if and only if M_1, M_2 have the same truth value for all first order sentences of quantifier depth k.*
2. *To each k-Ehrenfeucht value there is a first order sentence A of quantifier depth k such that the models M with that value are precisely those M for which A is true.*

2.6.2 The Simple Case of Total Order

The first order theory of total order, lets call it To, consists of equality and a single binary predicate $<$ satisfying the usual axioms

1. (Transitivity) If $x < y$ and $y < z$ then $x < z$.
2. (NonReflexivity) For no x is $x < x$.
3. (Trichotomy) For all x, y exactly one of $x = y$, $x < y$, $y < x$ holds.

We shall restrict our attention to *finite* models of To. There is a unique model for each finite size. For each positive integer n let $\overline{n}$ denote $\{1, \ldots, n\}$ with the usual order. We let $\overline{0}$ be the model with no points.

Now we analyze the k-move Ehrenfeucht Game on two finite models $\overline{n}, \overline{m}$. As described more generally in Subsection 2.6.1, we write $\overline{n} \equiv_k \overline{m}$ if Duplicator wins that game. To allow our induction to go we specify that for all $k \geq 1$ Duplicator wins the k-move Ehrenfeucht Game on two copies of $\overline{0}$ and loses the game on $\overline{0}, \overline{n}$ for $n \neq 0$.

Theorem 2.6.2. *Let $k \geq 1$. Then $\overline{n} \equiv_{k+1} \overline{m}$ if and only if*

1. *For all $1 \leq x \leq n$ there exists $1 \leq y \leq m$ such that $\overline{x-1} \equiv_k \overline{y-1}$ and $\overline{n-x} \equiv_k \overline{m-y}$*
2. *For all $1 \leq y \leq m$ there exists $1 \leq x \leq n$ such that $\overline{x-1} \equiv_k \overline{y-1}$ and $\overline{n-x} \equiv_k \overline{m-y}$*

Proof: Assume the conditions hold. Assume, by symmetry, that Spoiler's first move is some $x \in \overline{n}$. Duplicator plays the $y \in \overline{m}$ satisfying the condition. For the remaining k moves Duplicator plays as if there were separate games on the left boards $\{1, \ldots, x-1\}$, $\{1, \ldots, y-1\}$ and the right boards $\{x+1, \ldots, n\}$, $\{y+1, \ldots, n\}$. From the assumption Duplicator can win on both the left and the right boards. That is, $<$ is preserved for each pair of moves on the same side. But further, and critically, any two moves $a, b \in \overline{n}$ with $a < x < b$ have their corresponding moves $a', b' \in \overline{m}$ with $a' < y < b'$ so that $<$ is always preserved.

Now assume the conditions fail. Suppose, by symmetry, that there is an $x \in \overline{n}$ with no corresponding $y \in \overline{m}$. Spoiler selects that x and Duplicator must select some y. By symmetry suppose $\overline{x-1} \not\equiv_k \overline{y-1}$. Spoiler then makes his next k moves in $\overline{x-1}$ and $\overline{y-1}$. Duplicator must also move there, as all plays must be $<$ the first move. So Spoiler can win.

Note that models $\overline{n}$ of total order have a rather special property that a single move x splits the model into two separate parts. We shall see this behavior again when discussing strings in Chapter 10.

Theorem 2.6.3. *Let $k \geq 1$. Then $\overline{n} \equiv_k \overline{m}$ if and only if either $n = m$ or both $n, m \geq 2^k - 1$.*

Proof: We use induction on k. For $k = 1$ Spoiler need only make a play and so loses only when precisely one of n, m is zero. Assume the result for k and suppose the condition holds. If $n = m$ Duplicator clearly wins. Suppose $n, m \geq 2^{k+1} - 1$. By symmetry, assume Spoiler's first move is $x \in \overline{n}$. There are three cases.

1. (Low) $x \leq 2^k$. Duplicator selects $y = x \in \overline{m}$. Then $x - 1 = y - 1$ and both $n - x, m - y \geq 2^k - 1$ so $\overline{x-1} \equiv_k \overline{y-1}$ and $\overline{n-x} \equiv_k \overline{m-y}$ by induction.
2. (High) $n - x + 1 \leq 2^k$. Duplicator selects $y \in \overline{m}$ with $n - x + 1 = m - y + 1$. Then both $x - 1, y - 1 \geq 2^k - 1$ so $\overline{x-1} \equiv_k \overline{y-1}$ and $\overline{n-x} \equiv_k \overline{m-y}$ by induction.
3. (Middle) $x > 2^k$ and $n - x + 1 > 2^k$. Then both $x - 1, n - x + 1 \geq 2^k - 1$. Duplicator selects $y = 2^k \in \overline{m}$ so that $m - y \geq 2^{k+1} - 1 - 2^k = 2^k - 1$ and $y - 1 \geq 2^k - 1$ (in fact, equal). By induction both $\overline{x-1} \equiv_k \overline{y-1}$ and $\overline{n-x-1} \equiv_k \overline{m-y-1}$.

The conditions of Theorem 2.6.2 hold so $\overline{n} \equiv_k \overline{m}$.

Now suppose the conditions do not hold. First suppose $m \geq 2^{k+1} - 1$ but $n < 2^{k+1} - 1$. Spoiler plays $y = 2^k \in \overline{m}$. Then both $y - 1, m - y \geq 2^k - 1$. By Theorem 2.6.2 Duplicator would need play $x \in \overline{n}$ with $x - 1, n - x \geq 2^k - 1$ but, adding the two inequalities, that would imply $n \geq 2^{k-1} + 1$. Hence Spoiler wins. The case when $n \geq 2^{k+1} - 1$ and $m < 2^{k+1} - 1$ is symmetric. Finally, suppose both $n, m < 2^{k+1} - 1$ and $n \neq m$. At least one of them, say n, has $n < 2^{k+1} - 2$. Spoiler picks $x \in \overline{n}$ with $x - 1, n - x < 2^k - 1$. (He picks $x = 2^k - 1$ unless that's too big, in which case he simply picks $x = n$.) Duplicator must find $y \in \overline{m}$ with $\overline{x-1} \equiv_k \overline{y-1}$ and $\overline{n-x} \equiv_k \overline{m-y}$ which, by induction, would imply $x - 1 = y - 1$ and $n - x = m - y$ which would force $n = m$.

The formal inductive proof may mask somewhat the simple strategies. After u rounds in the Ehrenfeucht Game on $\overline{n}$, $\overline{m}$ the sets have been split by the moves into $u+1$ intervals, some possibly empty. With s moves remaining in the game Duplicator thinks of an interval as large if it has length at least $2^s - 1$. Her strategy is to make sure that after each round the corresponding intervals in the two models are all either exactly the same size or both large. Conversely, Spoiler's strategy is to make sure that this isn't the case: that there are a pair of corresponding intervals that are not the same size and are not both large.

Given two total orders M_1, M_2 we may naturally define their sum, we'll write it $M_1 + M_2$, by placing all of M_1 to the left of all of M_2. More formally, we say $x < y$ for all $x \in M_1, y \in M_2$ and when $x, y \in M_1$ or $x, y \in M_2$ then $x < y$ if and only if this was true for their own orders. As we shall see later with Theorem 10.2.2 the sum passes through to the $\equiv_k$ classes forming the Ehrenfeucht semigroup as given in Section 10.2.2 with Definition 10.2. Here, however, we can give the semigroup explicitly. The equivalences classes are given by $\overline{0}, \overline{1} \ldots, \overline{2^k - 2}$ together with the class containing $\overline{n}$ for all $n \geq 2^k - 1$. Let us use the symbol L (for large) for this last class and i for $\overline{i}$ for $0 \leq i \leq 2^k - 2$. Observe that $\overline{n} + \overline{m} = \overline{n+m}$. Thus the Ehrenfeucht Semigroup has elements $\{0, 1, \ldots, 2^k - 2, L\}$ and addition is normal addition except $L + y = y + L = L$ for all y and $i + j = L$ whenever normal addition gives an answer bigger than $2^k - 2$.

2.6.3 k-Similar Neighborhoods

We define the k-round *Distance Ehrenfeucht Game* on graphs G_1, G_2. As in our basic $\text{EHR}(G_1, G_2; k)$ Spoiler and Duplicator combine to select $x_1, \ldots,$ $x_k \in G_1$ and $y_1, \ldots, y_k \in G_2$. To win, in addition to the requirements of the basic Ehrenfeucht game, Duplicator must preserve distance. That is, using the definitions of Section 2.4, she must assure that $\rho(x_i, x_j) = \rho(y_i, y_j)$ for all $1 \leq i, j \leq k$. (For clarity, when one distance is infinite the other must also be.) We call this game $\text{DEHR}(G_1, G_2; k)$.

We are particularly interested in the Distance Ehrenfeucht Game when played on a neighborhood of a vertex.

Definition 2.9 *The d-neighborhoods of x, y are called k-similar if Duplicator wins the Distance Ehrenfeucht game on those neighborhoods that begins with x, y marked and has k additional rounds.*

We note that the notion of k-similarity is weaker than that of isomorphism and stronger than that of k-equivalence. We are also interested in the Distance Ehrenfeucht Game when played on the finite union of d-neighborhoods. Let $x_1, \ldots, x_u \in G_1$ and $y_1, \ldots, y_u \in G_2$. Call these the marked vertices. The d-picture of $x_1, \ldots, x_u$ (as defined in Section 2.4) splits into connected components $C_1, \ldots, C_r$. The d-picture of $y_1, \ldots, y_u$ splits into connected components $D_1, \ldots, D_{r'}$. Suppose $r = r'$. Suppose further that under suitable renumbering C_i, D_i contain corresponding marked vertices. That is, $x_l \in C_i$ if and only if $y_l \in D_i$.

Definition 2.10 *The d-pictures above are called* s-similar *if in addition to the above conditions for each corresponding pair of components C_i, D_i Duplicator wins the Distance Ehrenfeucht Game on C_i, D_i that begins with the $x_l \in C_i$, $y_l \in D_i$ marked and has s additional rounds.*

Property 2.6.4 *x_i, x_j lie in the same component of the d-picture of $x_1, \ldots,$ x_u if and only if there exists a sequence $i = i_0, \ldots, i_r = j$ so that the distance $\rho(x_{i_s}, x_{i_{s+1}}) \leq 2d + 1$ for $0 \leq s < r$.*

Proof: When this holds the d-neighborhoods of $x_{i_s}, x_{i_{s+1}}$ either have a common vertex or have adjacent vertices. In either case there is a path from x_{i_s} to $x_{i_{s+1}}$ and hence from x_i to x_j. Conversely suppose there is a path $x_i = z_0, \ldots, z_u = x_j$. For $0 \leq s \leq u$ let i_s be such that $\rho(x_{i_s}, z_s) \leq d$. There may be several choices for i_s, we only insist that we choose $i_0 = i$ and $i_u = j$. As z_s, z_{s+1} are adjacent the triangle inequality gives $\rho(x_{i_s}, x_{i_{s+1}}) \leq 2d+1$.

What happens when d is replaced by a smaller d'? Each component C_i may get smaller and, more important, it may split up into components $C_{i1}, C_{i2}, \ldots$. There can be no coalescing: x_i, x_j in separate components in the d-picture remain in separate components in the d'-picture.

Property 2.6.5 *Let $0 \leq d' < d$. Let $x_1, \ldots, x_u \in G_1$, $y_1, \ldots, y_u \in G_2$ have s-similar d-pictures. Then they have s-similar d'-pictures.*

Proof: From Property 2.6.4 two x_i, x_j lie in the same component in the d'-pictures if and only if there is a sequence of xs going from x_i to x_j with all distances at most $2d' + 1$. As the d-pictures are s-similar the distances between corresponding pairs of marked vertices must be the same. Hence the corresponding y_i, y_j would also lie in the same component of their d'-picture. Renumbering for convenience say $x_1, \ldots, x_l \in C_1$ and $y_1 \ldots, y_l \in C_2$ are the marked vertices in components C_1, C_2 in the d'-pictures. Let C_1^+, C_2^+ be the components containing these vertices in the d-pictures of G_1, G_2 respectively. By assumption Duplicator can win the Distance Ehrenfeucht Game on C_1^+, C_2^+ with s additional moves. Consider the Distance Ehrenfeucht Game on C_1, C_2. Duplicator plays with precisely the same strategy! That is, suppose Spoiler selects $x \in C_1$ at some stage. Duplicator plays the y she would have played in the C_1^+, C_2^+ game. This will certainly preserve equality, adjacency and distance. But is it a legal move? That is, is $y \in C_2$? We claim it is. Since $x \in C_1$ some $\rho(x, x_i) \leq d'$ with $1 \leq i \leq l$. As Duplicator has preserved distance to previously marked vertices $\rho(y, y_i) = \rho(x, x_i) \leq d'$. As y_i is a marked vertex in C_2, $y \in C_2$ as required.

Now we can give a powerful extension of Theorem 2.4.3

Theorem 2.6.6. *Set $d = \frac{3^k-1}{2}$. Suppose*

1. *For each* $y \in G_2$ *and each* $x_1, \ldots, x_{k-1} \in G_1$ *there is an* $x \in G_1$ *with* x, y *having* k*-similar* d*-neighborhoods and* $\rho(x, x_i) > 2d+1$ *for* $1 \leq i \leq k-1$.
2. *For each* $x \in G_1$ *and each* $y_1, \ldots, y_{k-1} \in G_2$ *there is a* $y \in G_2$ *with* x, y *having* k*-similar* d*-neighborhoods and* $\rho(y, y_i) > 2d+1$ *for* $1 \leq i \leq k-1$.

Then Duplicator wins $\text{EHR}(G_1, G_2; k)$.

Proof: We follow the template of Theorem 2.4.3. Set $d_0 = 0$ and define inductively $d_s = 3d_{s-1} + 1$ so that $d = d_s$ above. Suppose $x_1, \ldots, x_{k-s} \in G_1$, $y_1, \ldots, y_{k-s} \in G_2$ have been played. Duplicator's strategy is to have their d_s-neighborhoods s-similar. If she succeeds in holding to this strategy then at $s = 0$, the end of the Game, $d_s = 0$ and the 0-pictures are 0-similar. If x_i, x_j are adjacent they lie in the same component of their 0-picture and so the corresponding y_i, y_j also lie on the same component of their 0-picture. But 0-similarity preserves distance between marked vertices in the same component so y_i, y_j are adjacent and Duplicator has won. She can hold on to this strategy for the first move. Suppose (by symmetry) Spoiler selects $x \in G_1$ as his first move. From the assumption Duplicator can find $y \in G_2$ so that x, y have k-similar d-neighborhoods. Suppose inductively that she can hold to this strategy until there are s moves remaining. For convenience set $D = d_{s-1}$ so $d_s = 3D + 1$ and the $(3D+1)$-pictures are s-similar. Now (by symmetry) say Spoiler plays $x = x_{k-s+1} \in G_1$.

We call x inside if it is at distance at most $2D + 1$ from some previously marked vertex x_i, otherwise it is outside.

Suppose x is inside. Let $x \in C_1$, where C_1 is a component of the $(3D+1)$-picture of $x_1, \ldots, x_{k-s}$ and let C_2 be the corresponding component in G_2. For convenience let $x_1, \ldots, x_l$ denote the previously marked vertices of C_1 so that $y_1, \ldots, y_l$ denote the previously marked vertices of C_2. In the Distance Ehrenfeucht Game on C_1, C_2 Duplicator has a response y. Duplicator plays this y. As Duplicator is preserving distance inside components y is also inside. On C_1, C_2 the $(3D + 1)$-pictures of $x_1, \ldots, x_l, x$ and $y_1, \ldots, y_l, y$ are $(s-1)$-similar. From Property 2.6.5 on C_1, C_2 the D-pictures of $x_1, \ldots, x_l, x$ and $y_1, \ldots, y_l, y$ are $(s-1)$-similar. As x is inside the d-neighborhood of x in G_1 is contained in C_1. The D-picture of $x_1, \ldots, x_l, x$ in G_1 is therefore precisely the same as the D-picture of $x_1, \ldots, x_l, x$ in C_1. The same holds for y. Hence on G_1, G_2 the D-pictures of $x_1, \ldots, x_l, x$ and $y_1, \ldots, y_l, y$ are $(s-1)$-similar. The x_i and y_i that were in other components in the $(3D+1)$-pictures remain in s-similar, and hence $(s-1)$-similar, components in the D-pictures by Property 2.6.5.

Suppose x is outside. In the D-picture of marked $x_1, \ldots, x_{k-s}, x$ the D-neighborhood of x will be a distinct component. Duplicator moves to G_2 and selects y with x, y having k-similar, and hence $(s-1)$-similar D-neighborhoods. By the assumption she further chooses y far from previously marked vertices so that the D-neighborhood of y is a distinct component. The other components in the $(3D + 1)$-pictures remain s-similar, and hence $(s-1)$-similar, components in the D-pictures by Property 2.6.5.

We may also add the special vertices of Theorem 2.4.4. Let S_1, S_2 be subsets of G_1, G_2 respectively. We say that S_1, S_2 have k-similar d-neighborhoods if they have the same size and we may order their elements $s_1, \ldots, s_l \in S_1$, $s'_1, \ldots, s'_l \in S_2$ so that $s_1, \ldots, s_l$ and $s_1, \ldots, s'_l$ have similar k-neighborhoods in the sense of Definition 2.10. Note in particular that the induced graphs on S_1, S_2 must be isomorphic.

Theorem 2.6.7. *Set* $d = \frac{3^k-1}{2}$. *Suppose* G_1, G_2 *have vertex subsets* $S_1 \subset G_1, S_2 \subset G_2$ *with the following properties:*

1. S_1, S_2 *have* k*-similar* d*-neighborhoods as defined above.*
2. *Let* $d' \leq d$. *Let* $y \in G_2$ *with* $\rho(y, s_2) > 2d' + 1$ *for all* $s_2 \in S_2$. *Let* $x_1, \ldots, x_{k-1} \in G_1$. *Then there exists* $x \in G_1$ *with* x, y *having* k*-similar* d'*-neighborhood and such that* $\rho(x, x_i) > 2d' + 1$ *for* $1 \leq i \leq k-1$ *and* $\rho(x, s_1) > 2d' + 1$ *for all* $s_1 \in S_1$.
3. *Let* $d' \leq d$. *Let* $x \in G_1$ *with* $\rho(x, s_1) > 2d' + 1$ *for all* $s_1 \in S_1$. *Let* $y_1, \ldots, y_{k-1} \in G_2$. *Then there exists* $y \in G_1$ *with* y, x *having* k*-similar* d'*-neighborhood and such that* $\rho(y, y_i) > 2d' + 1$ *for* $1 \leq i \leq k-1$ *and* $\rho(y, s_2) > 2d' + 1$ *for all* $s_2 \in S_2$.

Then Duplicator wins $\text{EHR}(G_1, G_2; k)$.

The proof follows that of Theorem 2.6.6 much as the proof of Theorem 2.4.4 followed that of Theorem 2.4.3: Duplicator employs the strategy given in Theorem 2.6.6 with the vertices of S_1, S_2 marked at the start of the game.

Part II

Random Graphs

3. Very Sparse Graphs

We hold to the view proposed in the original papers of Erdős and Rényi that the random graph $G(n,p)$ *evolves* as p increases from empty to full. In its early stages – much like natural evolution – the behaviors are relatively simple to describe. For the random graph, early stages means up to $p \sim \frac{1}{n}$. As we are viewing the random graph through only a first order lens we shall actually go a bit further in this section. We summarize the results of Section 3.1 - 3.5 with Theorem 3.0.8.

Theorem 3.0.8. *Suppose $p(n)$ satisfies any of the following conditions.*

1. $p(n) \ll n^{-2}$
2. $n^{-1/k} \ll p(n) \ll n^{-1/(k+1)}$ *for some positive integer* k
3. $n^{-1-\epsilon} \ll p(n) \ll n^{-1}$ *for all positive* ϵ
4. $n^{-1} \ll p(n) \ll n^{-1} \ln n$
5. $n^{-1} \ln n \ll p(n) \ll n^{-1+\epsilon}$ *for all positive* ϵ

Then $p(n)$ satisfies the Zero One Law.

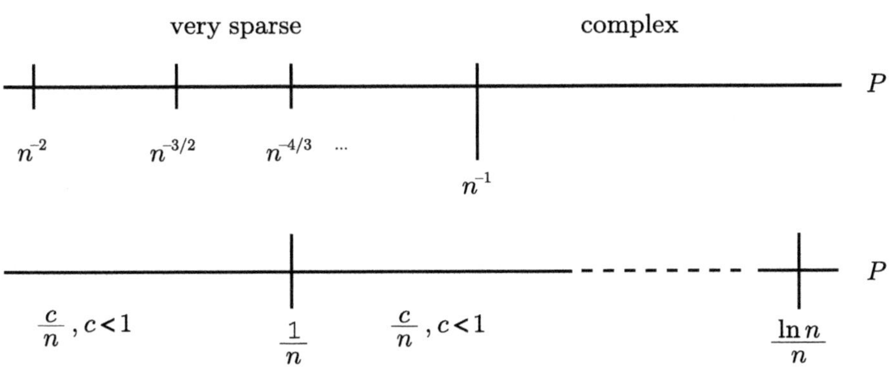

Fig. 3.1. The random graph evolves through different stages. When $p = n^{-c+o(1)}$ the behaviour for $c > 1$ is considerably simpler than that for $c < 1$. The case $c = 1$ deserves special attention: $p = \frac{1}{n}$ is the famous "double jump" of Erdős and Rényi, after which a giant component has emerged. Connectivity is achieved at $p = \frac{\ln n}{n}$

3.1 The Void

Suppose $p \ll n^{-2}$. Almost surely there are no edges. This makes things pretty simple.

Consider the Ehrenfeucht Game $\mathrm{EHR}[G_m, G_n; k]$ with $G_n \sim G(n, p(n))$ and $p(n) \ll n^{-2}$. Almost surely both G_m and G_n are empty. When $m, n \geq k$ and the graphs are empty there is nothing Spoiler can do. He selects a vertex on one graph and Duplicator selects one on the other. Duplicator wins. By the Bridge Theorem 2.5.1 there is a Zero-One Law.

What about the almost sure theory T? We have (as always) an axiom schema

$$P_r\text{: There exist (at least) } r \text{ vertices}$$

and the axiom

$$\textsc{NoEdge}: \forall_{x,y} \neg x \sim y$$

What are the countable models G for T? G must have a countable number of vertices and no edges. That determines G uniquely. T is $\aleph_0$-categorical and so is complete.

3.2 On the *k*-th Day ...

Here is a not so rough way to look at the early evolution. When p reaches $\Theta(n^{-2})$ edges appear. They remain scattered until p reaches $\Theta(n^{-3/2})$. Then edges with common vertices appear – which we'll consider as trees on three vertices. When p reaches $\Theta(n^{-1-1/k})$ trees on $k+1$ vertices appear. All of this occurs well before cycles appear at $\Theta(n^{-1})$.

Let $k \geq 1$ be integral and let $p = p(n)$ satisfy

$$n^{-1-\frac{1}{k}} \ll p(n) \ll n^{-1-\frac{1}{k-1}}$$

That is, $p(n)$ falls "between the cracks" of the appearances of trees on $k+1$ and $k+2$ vertices respectively. Almost surely

1. There are no cycles
2. There are no components with $k+2$ (or more) vertices
3. For all r every tree T on at most $k+1$ vertices appears as a component at least r times. This includes one vertex tree – there are at least r isolated vertices.

We outline the arguments. For the second property: There are $O(n^{k+2})$ choices of $k+2$ vertices and $O(1)$ (we're looking here at arbitrary but fixed k) choices of a tree on those vertices and probability p^{k+1} of having those tree edges and so the expected number of such configurations is $O(n^{k+2}p^{k+1})$ which is $o(1)$ by our upper bound on p. For the first property: For $l < k+2$

there are $O(n^l)$ choices of l vertices, $O(1)$ choices for a cycle, and probability p^l of having those cycle edges so the expected number of l-cycles is $O\big((np)^l\big)$ which is $o(1)$ as $p \ll n^{-1}$. For $l \geq k+2$ there are no cycles by the second property. For the third property fix $r \leq k+1$. There are $\Theta(n^r)$ choices of r vertices, $\Theta(1)$ (at least one) ways of placing a given tree T on those vertices, probability p^{r-1} of having the tree edges and probability more than $(1-p)^{r(n-1)}$ that there are no further edges involving those r vertices. As $p \ll n^{-1}$ the last factor is asymptotically one and so there are an expected number $\Theta(n^r p^{r-1})$ of tree components T. One can use the second moment method (or other techniques) to show that almost surely there are at least r components T for any fixed r.

The three properties give a first order theory T. What are the countable models G of T? Each component can have at most $k+1$ vertices and must be a tree, including possibly an isolated vertex. But every tree does occur as a component at least r times for all r. So G must consist of a countable number of copies of every tree on at most $k+1$ vertices and nothing else. That is: G is uniquely determined and so T is $\aleph_0$-categorical and hence there is a Zero-One Law.

3.3 On Day ω

As p evolves it runs through a countable sequence of threshold functions, $n^{-1-1/k}$ for $k = 1, 2, \ldots$. Now we examine the "gap" between those functions and the next (and arguably most important) threshold function, n^{-1}. We suppose here that $p(n) \gg n^{-1-\epsilon}$ for all positive ϵ but $p(n) \ll n^{-1}$. This would include such functions as $p(n) = 1/[n \ln n]$. For these $p(n)$, the following properties hold almost surely:

- There are no cycles
- For all r, every finite tree T occurs at least r times as a component.

Consider, as in the previous sections, the countable models G of this theory T. Every finite tree T occurs as a component an infinite number of times. All components, as there are no cycles, must be trees. But now a new phenomenon occurs. There may or may not be infinite tree components. Our object, now, is to show that such infinite tree components do not matter from a first order perspective. Roughly, infinite tree components will be simulated by appropriate large trees.

3.3.1 An Excursion into Rooted Trees

Consider a rooted tree, by which we mean a tree T (finite or infinite) and a designated vertex $R \in T$, called the root. With rooted trees the terms parent, child, ancestor and descendant all have their natural computer science meaning. The *depth* of a vertex v is its distance from the root R. For any

$w \in T$ we let T^w be the subtree consisting of w and all of its descendants. We define the (r, s)-value of a rooted tree for positive integers r, s. Roughly, we examine the r-neighborhood of the root v and consider any count greater than s as "many". For $r = 1$ the (r, s)-values are indeed $0, 1, \ldots, s$ and M for many. The value is the number of children of v, with M if that number is bigger than s, including infinite. Now suppose by induction that the (r, s)-value has been defined and let $\text{VAL}(r, s)$ denote the set of such values. For each child w of the root v consider the (r, s)-value σ_w of T^w. For each $\sigma \in \text{VAL}(r, s)$ let $u(\sigma)$ denote the number of children w of v with $\sigma_w = \sigma$. However, if this number is bigger than s, including infinite, call it M, for many. Then u defines a function $u\colon \text{VAL}(r, s) \to \{0, \ldots, s, M\}$. This function is the $(r + 1, s)$-value of the original rooted tree.

It might be helpful to think of the (r, s)-value computationally. Consider the tree T with root v and cut off the tree at the r-th generation. Now we give each vertex a value working from the r-th generation up. The vertices in the r-th generation all have the same value, call it O. The vertices in the $(r-1)$-st generation have values $0, 1, \ldots, s, M$, their number of children. The values of the vertices in the $(r - 2)$-nd generation are functions $u\colon \{0, 1, \ldots, s, M\} \to \{0, 1, \ldots, s, M\}$, meaning they have $u(i)$ children with i children. We continue up the tree until reaching v and the value given it is its (r, s)-value. Note that the number of (r, s)-values is a tower of $(s + 2)$'s of height r, a very large but finite quantity.

Examples: Let $r = 2, s = 3$. Values for the rooted tree include

1. No children
2. Two children with many children and many children with one child and no children with zero, two or three children.
3. One child with no child, three children with one child, many children with two children, no children with three children and two children with many children.

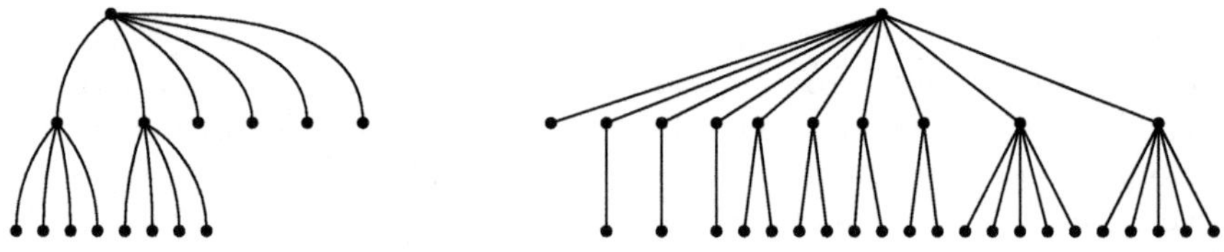

Fig. 3.2a.

Each of these (r, s)-values can be expressed in a first order language involving the root as constant symbol, and a binary parent/child relation. Taking disjuncts of such expressions leads to an interesting class of properties. Consider, just for fun, your maternal grandmother. Let many mean, as above, more than three. Is it true or false that

1. She had many children
2. She had children with many children

3. (Author's favorite!) She had no children that had no children that had no children.

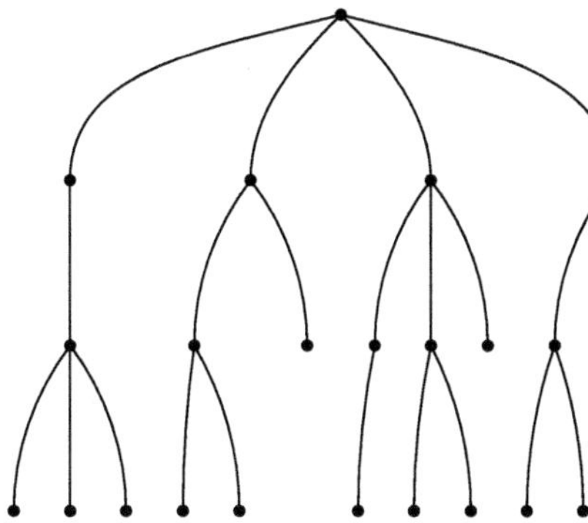

Fig. 3.2b. Did the author's grandmother have no children that had no children that had no children?

Let $\sigma \in \text{VAL}(r, s)$. We can easily create a finite tree T with designated root v which has value σ. We simply interpret "many" as $s+1$. Now the tree down to the r-th generation is determined and we halt it there.

We may naturally speak of considering a rooted tree T as a graph. To do so we remove the special designation of the root and consider v, w adjacent if either v is the parent of w or w is the parent of v.

Theorem 3.3.1. *Let T_1, T_2 be rooted trees with roots R_1, R_2 which have the same $(r, s-1)$-value. Then, with T_1, T_2 considered as graphs, R_1, R_2 have s-similar r-neighborhoods in the sense of Definition 2.9.*

Proof: Let $R_1 = x_0, x_1, \ldots, x_s \in T_1$ and $R_2 = y_0, y_1, \ldots, y_s \in T_2$ be the marked vertices in the Ehrenfeucht Game on the r-neighborhoods beginning with R_1, R_2 marked and having s additional rounds. To show Theorem 3.3.1 we make a stronger induction hypothesis.
Induction Hypothesis (on r): Duplicator can play to insure that

- $x_i = x_j$ if and only if $y_i = y_j$
- $\rho(x_i, x_j) = \rho(y_i, y_j)$. In particular, with $i = 0$, x_j, y_j have equal depth.
- x_i is an ancestor of x_j if and only if y_i is an ancestor of y_j.

For $r = 0$ this is trivial as Spoiler and Duplicator must repeatedly choose the roots. Suppose $r = 1$. If the roots have the same number of children then the 1-neighborhoods are isomorphic and Duplicator wins. Otherwise both roots have at least s children and Duplicator wins by matching children for s rounds.

Assume the result for r and let T_1, T_2 have the same $(r+1, s-1)$-value. Call $x^* \in T_1$ a *chieftain* if it is a child of the root. Each nonroot $x \in T_1$ has a unique chieftain x^* with $x \in T^{x^*}$ (possibly $x^* = x$), we say x^* is the chieftain of x. We use the same notation for T_2. Now suppose Spoiler plays, in some round, $x \in T_1$. If x is the root Duplicator simply responds by playing the root of T_2. Otherwise let x^* be the chieftain of x. Duplicator gives Spoiler a free move and imagines he has played x^* and x.

She first responds to x^*. If x^* had already been played there already is a corresponding y^*. Otherwise, let α denote the $(r, s+1)$-type of x^*. Duplicator finds a chieftain $y^* \in T_2$ who also has type α and has not yet been played. We use the $r = 1$ argument to find y^*. Either T_1, T_2 have the same number of chieftains of type α or they both have at least s such chieftains. In the first case Duplicator matches the chieftains. In the second case Duplicator can match the chieftains for s rounds since even with the fictional moves only s chieftains are being selected in each tree.

Now suppose chieftains x^*, y^* are matched and both have the same $(r, s+1)$-value. By induction Duplicator can respond on T^{x^*}, T^{y^*} preserving equality, ancestor and distance. Note that she needs do this for at most s plays, once the chieftains have been marked.

Has Duplicator won? Suppose x_i is an ancestor of x_j. If they have a common chieftain x^* then y_i, y_j have a common chieftain y^* and y_i is an ancestor of y_j by the induction hypothesis. Otherwise x_i is the root but then y_i is also the root so y_i is again an ancestor of y_j.

What about depth? If x_i has depth zero it is the root so y_i has depth zero. Otherwise let x^*, y^* be their chieftains. By induction, if u is the depth of x_i in T^{x^*} then u is the depth of y_i in T^{y^*}. Then x_i, y_i have depth $u+1$ in T_1, T_2 respectively.

What about distance? If x_i, x_j have a common chieftain x^* then their distance in T_1^* is their distance in T^{x^*}. By induction y_i, y_j have the same distance. If x_i is the root then so is y_i and distance is depth which is preserved. Finally, suppose x_i, x_j do not have a common chieftain. Any path from x_i to x_j must go through the root. Hence the distance $\rho(x_i, x_j)$ is the sum of the depth of x_i and the depth of x_j. As depth is preserved so is distance.

The inductive proof of Theorem 3.3.1 somewhat masks an algorithmic implementation for Duplicator. When Spoiler plays $x \in T_1$ Duplicator considers the entire path from the root to x as being played. Duplicator starts down that path marking corresponding vertices in T_2. She makes sure that if x, y are corresponding vertices which are both at depth i then they both have the same $(r-i, s-1)$-value. Some initial segment of that path (certainly the root itself) already has corresponding vertices. If all do there is no problem. Otherwise there is an x^+, x^- on the path at depths $i, i+1$ where x^+ already has a corresponding y^+ but x^- does not. Let α be the $(r-i-1, s-1)$-value of x^-. If x^+, y^+ have the same number of children with that value then Duplicator finds a corresponding y^-. If x^+, y^+ both have at least s children with that value then Duplicator can find a corresponding y^- since at most $s-1$ have been marked up to this point. Duplicator then continues down the path in T_1 finding corresponding vertices in T_2 of the same type.

3.3.2 Two Consequences

Theorem 3.3.2. *Let G_1, G_2 both be acyclic graphs in which every finite tree occurs as a component an infinite number of times. (G_1, G_2 may or may not have infinite tree components.) Then G_1 and G_2 are elementarily equivalent.*

Proof: Fix k and set $d = \frac{3^k-1}{2}$. Consider the d-neighborhood of a vertex x in, say, G_1. This may be considered as a rooted tree with x as the root. As such it has a $(d, k-1)$-value α. For every value α there is a finite rooted tree T with that value. G_2 contains an infinite number of components isomorphic to T. For each such component the d-neighborhood of its root y is k-similar to the d-neighborhood of x by Theorem 3.3.1. Given any $y_1, \ldots, y_{k-1} \in G_2$ we could find such a $y \in G_2$ lying in a different component. The assumptions of Theorem 2.6.6 are therefore satisfied and so $G_1 \equiv_k G_2$.

As all countable models of T are elementarily equivalent T is itself complete and so the Zero-One Law holds for those $p = p(n)$ with $p(n) \ll n^{-1}$ and $p(n) \gg n^{-1-\epsilon}$ for all positive ϵ.

Theorem 3.3.3. *For $s \geq 3$, s-Colorability is not first order expressible.*

Proof: We use a beautiful theorem of Erdős [4]: there exist finite graphs of arbitrarily high chromatic number and, simultaneously, arbitrarily high girth. We shall use a bit more, that for all s, a, b, c, D there exists a graph G such that

1. G is not s-colorable
2. G has no cycles of length $\leq a$.
3. All vertices of G have degree at least b
4. Given any c vertices of G there is another vertex which is not within distance D of any of them.

These additional properties follow easily from the proof of Erdős. Alternatively, take a graph G which is not s colorable and with no cycles of length $\leq a$. Make each vertex of G the root of an infinite tree (all on disjoint vertex sets) where every vertex has infinite degree. Then the final two properties hold for all b, c, D.

Let k be arbitrary. Set $d = \frac{3^k-1}{2}$. Let G_1 satisfy the above with $a = 2d+1$, $b = k+1$, $c = k-1$ and $D = 2d+1$. Let G_2 be the infinite rooted tree with all vertices of infinite degree so that G_2 is s-colorable, indeed 2-colorable.

Consider the d-neighborhood of any $x \in G_1$. As no cycles have size at most $2d+1$ this is a rooted tree. In this rooted tree every parent will have at least k children. (Of the at least $k+1$ adjacencies only one goes to a parent.) The $(d, k-1)$-value is thus determined. (For example: When $d = 2$, x has many children with many children and no others.) The d-neighborhoods of the $y \in G_2$ are also rooted trees and have the same $(d, k-1)$-value. Given any $x_1, \ldots, x_{k-1} \in G_1$ one can find $x \in G_1$ with all $\rho(x, x_i) > 2d+1$. In the infinite rooted tree G_2 this is certainly true. The assumptions of Theorem 2.6.6 therefore apply and so $G_1 \equiv_k G_2$. Theorem 3.3.3 then follows by the general Theorem 2.3.3.

We have actually shown a somewhat stronger nonseparability result: Fix $s \geq 3$. There is no first order sentence A which holds for all graphs which have chromatic number two and fails for all graphs with have chromatic number larger than s.

3.4 Past the Double Jump

For the random graph theorist the random graph $G(n,p)$ undergoes a critical transition – a percolation – at $p \sim \frac{1}{n}$. The bulk of this transition is not visible through our first order lens. Now we look just past this transition. In this section we assume

$$\frac{1}{n} \ll p(n) \ll \frac{\ln n}{n} \tag{8}$$

At this stage cycles have appeared but more complex local subgraphs have not yet appeared. There are still vertices of small degree but they do not occur near the cycles. The almost sure theory can be describe by four schema.

1. For every k: There are no k vertices with (at least) $k+1$ edges.
2. For every r and every $k \geq 3$: There are (at least) r cycles of size k.
3. For every s, d and every $k \geq 3$: There does not exist a cycle of size k and a vertex of degree d at distance s from the cycle. (We include the case $s = 0$: No vertex of a k-cycle can have degree d.)
4. For every r and every finite tree T: There are (at least) r components isomorphic to T.

We outline the arguments. Set $\alpha = pn$ for convenience. There are $O(n^k)$ choices for k vertices, $O(1)$ choices for $k+1$ edges (critically, k is fixed) and probability p^{k+1} of having those edges and $O(n^k p^{k+1}) = O(\alpha^{k+1} n^{-1}) = o(1)$. For the second property we note there are $(n)_k/2k = \Theta(n^k)$ potential k-cycles and each has all k edges in $G(n,p)$ with probability p^k. Thus the expected number of k-cycles is $\Theta\big((np)^k\big)$ which approaches infinity. A second moment method shows that almost surely the number is close to its expectation. For the third property: There are $O(n^{k+s+d-1})$ choices for $k+s+d-1$ vertices, $O(1)$ ways to place a k-cycle, a path of length s from it to a vertex v, and $d-1$ further neighbors of v. There is probability $p^{k+s+d-1}$ of having those edges. But there is also probability $(1-p)^{n-1-d}$ that v is not adjacent to any other vertices. Setting $k+s+d-1 = u$ there are an expected number $O(\alpha^u e^{-\alpha})$ of such configurations and as $\alpha \to \infty$ this is $o(1)$. For the final property fix a tree T with k vertices. There are $O(n^k)$ choices for the k vertices, $O(1)$ choices for placing T, probability p^{k-1} of having the edges and probability roughly $(1-p)^{nk}$ that the vertices have no further adjacencies.

The expected number of such T is then $\sim n\alpha^{k-1}e^{-k\alpha}$. Now we use the upper bound on $p(n)$. As $\alpha = o(\ln n)$ the n factor dominates, the expected number of T goes to infinity. Again, a second moment method shows that almost surely the number of T-components is almost surely close to its expectation.

Let T be the theory generated by the above four schema and let G be a countable model of T. The first schema rules out bicyclic (or more) components so all components are either trees or unicyclic. (It is interesting to compare the countable models with the finite graphs $G\big(n, p(n)\big)$: For $\epsilon > 0$ arbitrarily small and $p > (1+\epsilon)/n$ there is a giant component with v vertices and e edges where e is substantially bigger than $v+1$. However, this is not "seen" through our first order lens.) Suppose a component contains a k-cycle. From the axiom schema there can be no more cycles and there can be no vertices of finite degree. This uniquely determines the component, from each vertex of the k-cycle there rises an infinite tree with each vertex having an infinite number of children. G must have a countable number of such components for each k. Also, G must have a countable number of T-components for every finite tree T. What is left open is that G may or may not have infinite trees, and if so they can be of any variety. Now let G_1, G_2 be two countable models. From Theorem 3.3.2 Duplicator can win $\textsc{Ehr}(G_1, G_2; k)$ if one throws out the components containing cycles. But the components containing cycles are precisely the same in both graphs so that it doesn't pay Spoiler to play there. Thus Duplicator wins $\textsc{Ehr}(G_1, G_2; k)$ for any k, so that G_1, G_2 are elementarily equivalent. This implies that the theory T is complete. Finally, that implies a Zero-One Law for these $p(n)$.

3.5 Beyond Connectivity

At $p \sim \frac{\ln n}{n}$ the random graph $G(n,p)$ becomes connected. In the first order world it loses all its isolated vertices, indeed, all its vertices of low degree. Now we assume

$$\frac{\ln n}{n} \ll p(n) \ll n^{-1+\epsilon} \tag{9}$$

for all $\epsilon > 0$. The almost sure theory T is given by three schema:

1. For every k: There do not exist k vertices with (at least) $k+1$ edges.
2. For every d: All vertices have (at least) d neighbors
3. For every r and every $k \geq 3$: There exist (at least) r cycles of size k.

The first and third properties are shown as in the previous section. For the second the expected number of vertices of degree precisely i is $n\binom{n-1}{i}p^i(1-p)^{n-1-i}$ which is roughly $n\,(np)^i e^{-np}$. With $np \gg \ln n$ (which could actually be relaxed somewhat) the e^{-np} term dominates and this is $o(1)$.

What about the countable models G of this theory T? By the second schema all vertices must have infinite degree. All components must be trees or unicyclic. For every $t \geq 3$ there are an infinite number of t-cycles and each one is in a component where every vertex of the t-cycle is the root of a tree where all vertices have an infinite number of children. The tree components of G must have all vertices of infinite degree and so are determined up to isomorphism. But T is not $\aleph_0$-categorical as G may or may not have such tree components.

Let G_1, G_2 be two models and let k be fixed. Set $d = \frac{3^k-1}{2}$. Let $S_1 \subset G_1$, $S_2 \subset G_2$ be those vertices in small cycles. We now apply Theorem 2.4.4. As S_1, S_2 both consist of an infinite number of cycles of each size $t \geq 3$ we have an isomorphism between them. When $y \in G_2$ lies more than $2d'+1$ from S_2 its d'-neighborhood is determined as a rooted tree when each parent has an infinite number of children. Given any $x_1, \ldots, x_{k-1} \in G_1$ we can find $x \in G_1$ that is more than $2d'+1$ from any cycle and from any x_i. Then y, x have isomorphic d'-neighborhoods. We deduce $G_1 \equiv_k G_2$. As k was arbitrary the models are elementarily equivalent, and thus we have a Zero-One Law for these $p = p(n)$.

Notice a curious fact. In this range the random graph $G(n,p)$ is almost surely connected. The countable models of the almost sure theory T in this range are definitely not connected. There is no contradiction as connectivity is not a first order property. Indeed, this may be viewed as a most roundabout way of proving that connectivity is not a first order property!

3.6 Limiting Probabilities

At threshold functions we generally do not have a Zero-One Law – there are first order sentences whose probability is "moving" between zero and one and has some intermediate value. In some cases we are able to describe those intermediate values. Generally this occurs when the almost sure theory T is almost complete. We begin with a general result and then apply it to various threshold functions.

3.6.1 A General Result on Limiting Probabilities

Suppose that for every $n \geq 2$ we have a probability space over the graphs on n vertices which we label G_n. (In our applications we will always have $G_n \sim G\big(n, p(n)\big)$.) Let T be the almost sure theory – i.e., the set of sentences A with $\lim_{n\to\infty} Pr\big[G_n \models A\big] = 1$. Let I be a countable index set and let σ_i, $i \in I$ be a set of sentences not in T.

Definition 3.1 *A family σ_i is a* complete set of completions *(relative to theory T) if*

1. *$T + \sigma_i$ is complete for all $i \in I$*
2. *For all distinct $i, j \in I$, $T \models \neg(\sigma_i \wedge \sigma_j)$*
3. *For each $i \in I$, $\lim_{n\to\infty} \Pr\big[G_n \models \sigma_i\big]$ exists.*

4. $\sum_{i \in I} \lim_{n\to\infty} \Pr\left[G_n \models \sigma_i\right] = 1$

We will set $p_i := \lim_{n\to\infty} \Pr\left[G_n \models \sigma_i\right]$ for notational convenience. For any first order A let $S(A)$ denote the set of $i \in I$ for which $T + \sigma_i \models A$.

Theorem 3.6.1. *Under the above conditions* $\lim_{n\to\infty} \Pr\left[G_n \models A\right]$ *exists for all first order* A *and is given by*

$$\lim_{n\to\infty} \Pr\left[G_n \models A\right] = \sum_{i \in S(A)} p_i \tag{10}$$

In the examples later we shall give explicit σ_i with calculatable p_i for which the family of $S(A)$ is completely describable. This will allow us to give a complete description of the possible values of $\lim_{n\to\infty} \Pr\left[G_n \models A\right]$.

Proof: Let $\epsilon > 0$ be arbitrarily small and set $P = \sum_{i \in S(A)} p_i$. We shall show $\Pr\left[G_n \models A\right] > P - \epsilon$ for n sufficiently large. Pick a *finite* set $J \subseteq S(A)$ with $\sum_{i \in J} p_i > P - \frac{\epsilon}{2}$. Set $\sigma := \bigvee_{i \in J} \sigma_i$. Then $T + \sigma \models A$. By compactness (the finiteness of proof) one only needs a finite number of sentences of T above and so there is a single $\tau \in T$ (taking the conjunction of the sentences used) so that $\tau \wedge \sigma \models A$. As τ has limiting probability one (being in T) and σ has limiting probability at least $P - \frac{\epsilon}{2}$, for n sufficiently large $\sigma \wedge \tau$ has probability at least $P - \epsilon$, but then A has at least that probability.

Now consider $\neg A$. As all $T + \sigma_i$ are complete, $S(\neg A) = \overline{S(A)}$ and as $\sum_I p_i = 1$, $\sum_{i \in S(\neg A)} = 1 - P$. Applying the above argument to $\neg A$, for n sufficiently large $\Pr\left[G_n \models \neg A\right] > 1 - P - \epsilon$ which gives an upper bound $\Pr\left[G_n \models A\right] < P + \epsilon$.

3.6.2 In the Beginning ...

Here we consider the parametrization $p = c/\binom{n}{2}$ with c a positive constant. Edges may or may not have appeared.

The almost sure theory T includes (indeed, is generated by):

1. For all r: There are (at least) r isolated vertices.
2. For no distinct x, y, z are both x, y adjacent and x, z adjacent. That is, all edges are isolated components.

Let I be the nonnegative integers and let σ_i be the sentence: There exist precisely i edges.

We first claim $T + \sigma_i$ is complete. A countable model G_i of $T + \sigma_i$ must consist of a countable number of isolated vertices and precisely i isolated edges and nothing else. This makes $T + \sigma_i$ $\aleph_0$-categorical and hence complete. Tautologically $\neg(\sigma_i \wedge \sigma_j)$ for all $i \neq j$. The number of edges in $G(n,p)$ is given precisely by a binomial distribution which is asymptotically the Poisson distribution with mean c. Thus $p_i = e^{-c}\frac{c^i}{i!}$ is the limiting probability of σ_i. Finally $\sum p_i = 1$. Thus the σ_i do form a complete set of completions.

What are the possible $S(A)$? Let k be such that $G \equiv_k G'$ implies that G, G' agree on A. Consider $\text{EHR}(G_i, G_j; k)$ where G_i, G_j are the countable models described above. When $i, j \geq k$ Duplicator wins – every time Spoiler selects a vertex on a new edge in one graph she does the same on the other graph. Thus either $i, j \in S(A)$ or $i, j \notin S(A)$. This forces $S(A)$ to be finite or cofinite. Further, any finite set S is $S(A)$ for the sentence $\bigvee_{i \in S} \sigma_i$ and any cofinite set S is $S(A)$ for the sentence $\neg \bigvee_{i \notin S} \sigma_i$. We apply Theorem 3.6.1 and conclude:

Theorem 3.6.2. *Set $p = c/\binom{n}{2}$. For every first order A*

$$\lim_{n \to \infty} \Pr\left[G(n,p) \models A\right]$$

exists. Furthermore it is either a finite sum of terms $e^{-c}c^i/i!$ (including zero as the empty sum) or one minus such a finite sum. Moreover, for every such sum there is an A with that limiting probability.

3.6.3 On the k-th Day ...

Here we let k be a fixed positive integer and parametrize $p = cn^{-1-1/k}$. (This generalizes the case $k = 1$ done in the previous section.) At this stage trees with $k+1$ vertices are appearing. The almost sure theory T is given by the schema:

1. For every tree T on fewer than $k+1$ vertices (including the one point "tree") and every r: There exist (at least) r components T
2. There do not exist $k+2$ vertices whose edges include a tree.

Let $T_1, \ldots, T_u$ denote the possible trees (up to isomorphism) on $k+1$ vertices. [Expressing u as a function of $k+1$ is an intriguing problem in asymptotic combinatorics but for our purposes here it is simply a positive integer.] Let I denote the set of u-tuples $\mathbf{m} = (m_1, \ldots, m_u)$ of positive integers. We let $\sigma_{\mathbf{m}}$ be the sentence: There exist (precisely) m_i components T_i for $1 \leq i \leq u$.

Theories $T + \sigma_{\mathbf{m}}$ are complete as they are $\aleph_0$-categorical. The models $G_{\mathbf{m}}$ of $T + \sigma_{\mathbf{m}}$ have countably many copies of every tree component of size less than $k+1$, precisely those components of size $k+1$ given by $\mathbf{m}$, and nothing else. Tautologically no two of the $\sigma_{\mathbf{m}}$ can hold simultaneously. The number of components T_i is asymptotically Poisson with mean $\mu_i := \lambda_i c^k$, where λ_i depends on the number of automorphisms of T_i. These numbers are asymptotically independent so that $p_{\mathbf{m}}$ is given by the product over i of $e^{-\mu_i}\mu_i^{m_i}/m_i!$ The sum over all $\mathbf{m}$ breaks into the product of u terms, each of which is one.

Fix s and ask whether $G_{\mathbf{m}} \equiv_s G_{\mathbf{m}'}$. If for each i either the coefficients m_i, m_i' are equal or they are both at least s then Duplicator does win $\text{EHR}(G_{\mathbf{m}}, G_{\mathbf{m}'}; s)$ as Spoiler has not enough rounds to take advantage of the distinctions. This allows a full description of the possible $S(A)$. Call an s-cylinder the set of $\mathbf{m}$ where each coordinate is either a constant less than s or any integer s or bigger. Then $S(A)$ is the finite union of such cylinders (for some s). Conversely, it is easy to construct A giving any $S(A)$ of this form.

The final result is a bit messy to write down but it is precise – there is an exact characterization of the limit probabilities of first order A. It is easier to step back and note that those limit probabilities, as functions of c, have a relatively nice form. They are products of terms which are sums or one minus sums of terms which themselves are powers of c times exponentials of the k-th power of c. One simple but powerful corollary: They are continuous functions of c.

3.6.4 At the Threshold of Connectivity

Perhaps the most stunning result in the original papers of Erdős and Rényi on random graphs was their threshold for connectivity. They showed that if

$$p = \frac{\ln n}{n} + \frac{c}{n}$$

with c any real constant then

$$\lim_{n\to\infty} \Pr\big[G(n,p) \text{ connected}\big] = e^{-e^{-c}}$$

One can split their proof into two parts. In the first part they show that with p as above the limit probability that $G(n,p)$ has no isolated vertices is $\exp[-e^{-c}]$ as above. In the second part they showed that with p as above the events "G connected" and "G has no isolated vertices" are almost surely the same. In our first order world it is the isolated vertices that play the key role. We shall show that any first order A is, up to $o(1)$ probability, equivalent to making a rather restricted statement about the number of isolated vertices. The results of this subsection are due to Lubos Thoma and the author [22].

The almost sure theory T for p in this range (for any c) is given by the axiom schema:

1. For every k: There are no k vertices with $k+1$ edges
2. For every $k \geq 3$ and every r: There are (at least) r k-cycles.
3. For every $k \geq 1$ and every r: There are (at least) r vertices with (precisely) k neighbors.
4. For every $k, k' \geq 3$ and every d: There do not exist a k-cycle and k'-cycle at distance d.
5. For every $k \geq 3$, $k' \geq 1$ and every d: There do not exist a k-cycle and a vertex of degree k' at distance d.
6. For every $k, k' \geq 1$ and every d: There do not exist vertices v, v' of degrees k, k' at distance d.

The fourth and fifth schema include $d = 0$, no two cycles can intersect and no cycle can have a vertex of fixed degree. What are the countable models G of T? The cycles and vertices of finite degree must all lie in separate components by the last three schema. Every k-cycle is thus in a uniquely determined component, in which each vertex of the cycle is the root of an infinite tree in which every vertex has an infinite number of children. Every vertex of degree k' is in a uniquely determined component, a tree in which all other vertices have an infinite number of children. There are a countable number of such components for every $k \geq 3$ and all $k' \geq 1$. There may or may not be trees in which every vertex has an infinite number of children. Finally, the threshold property, there may or may not be isolated vertices.

Let I be the nonnegative integers and let σ_i be the sentence "There exist precisely i isolated vertices." We claim these form a complete set of completions. The countable models of $T+\sigma_i$ are as above except that their numbers of isolated vertices are fixed. Any two models G_1, G_2 of $T + \sigma_i$ differ only in that one may contain trees with every vertex having infinite degree. Split $G_1 = G_1^t \cup G_1^u$ where G_1^t consists of the tree components and G_1^u consists of the unicyclic components. Similarly split $G_2 = G_2^t \cup G_2^u$. For any k $G_1^t \equiv_k G_2^k$ by Theorem 3.3.2 and $G_2^u \equiv_k G_2^u$ since they are isomorphic. Hence $G_1 \equiv_k G_2$. That is, G_1, G_2 are elementarily equivalent. Thus $T + \sigma_i$ is complete. Two distinct σ_i tautologically cannot simultaneously hold. From classical results the number of isolated vertices in $G(n,p)$ behaves asymptotically like the Poisson distribution with mean $\mu := e^{-c}$. Thus $\Pr[\sigma_i]$ approaches $e^{-\mu}\mu^i/i!$ which sums to one.

Let G_i, G_j be countable models of $T + \sigma_i$ and $T + \sigma_j$. If $i, j \geq k$ then Duplicator wins $\text{EHR}(G_i, G_j; k)$ as Spoiler does not have sufficiently many rounds to take advantage of the difference in the number of isolated vertices. Thus for every A there is a k so that either all or none of the $i \geq k$ are in $S(A)$. In other words, $S(A)$ is finite or cofinite. Conversely (as we saw with $p = c/\binom{n}{2}$ earlier) all finite and cofinite sets are possible. We deduce:

Theorem 3.6.3. *Let $p = \frac{\ln n}{n} + \frac{c}{n}$ and set $\mu := e^{-c}$. Then for every first order A, $\lim_{n\to\infty} \Pr\left[G(n,p) \models A\right]$ exists and is a finite sum of terms of the form $e^{-\mu}\mu^i/i!$ or one minus such a sum. Further, all such expressions do occur for some first order A.*

3.7 The Double Jump in the First Order World

Perhaps the most important discovery of Erdős and Rényi in their investigation of the evolution of the random graph was the global change in the nature of $G(n,p)$ near $p = 1/n$. When $p = (1-\epsilon)/n$ (ϵ arbitrarily small but fixed positive) then the components of G are all small, the largest has size $O(\ln n)$. The components are all trees or unicyclic. But when $p = (1+\epsilon)/n$ a giant component, with $c_\epsilon n$ vertices, has emerged. This component has far more edges than vertices. The remaining vertices lie in small components, of size $O(\ln n)$, which are all trees or unicyclic. At $p = 1/n$ the situation is in between, the largest components have size $\Theta(n^{2/3})$. The existence of components that were neither trees nor unicyclic was open- it had limiting probability strictly between zero and one. Erdős and Rényi dubbed this phenomenon the double jump. Today it is often thought of as a percolation phenomenon.

In this section we examine the almost sure theory and limiting probabilities when $p = \lambda/n$, for λ an arbitrary positive constant. We are able to give a complete description of those limiting probabilities as functions of λ. The most interesting result is in the negative: there is nothing special about $\lambda = 1$ for these limiting probabilities. That is, in our first order world the double jump is not seen!

3.7.1 Poisson Childbearing

Let P_λ denote the Poisson Distribution with mean λ. That is:

$$\Pr[P_\lambda = i] = e^{-\lambda}\frac{\lambda^i}{i!}$$

Start with Eve. She has P_λ children. They (conveniently, all female!) in turn have P_λ children and those children in turn beget children. Stop the procedure at the end of the r-th generation. [It may have already stopped if, for example, Eve was childless.] The random rooted (at Eve) tree T thus formed we call the (r, λ) Poisson tree.

This is an example of what probabilists know as a Galton-Watson process. In their studies they allow the tree to go on forever [no limit on generations] and study such questions as whether or not the tree is finite. We shall be content to stop the process at an arbitrary, but fixed, generation r.

Now we seemingly switch gears and study the r-neighborhood in $G(n, \lambda/n)$ of a given vertex v. The number of neighbors of v is given by a Binomial Distribution which is asymptotically P_λ. Now these neighbors have new neighbors with asymptotic distribution P_λ. This continues down to the r-th level. Actually, there are a number of technical problems here. A vertex could be adjacent to another vertex already in the tree, forming a cycle. Also, if the tree currently has u vertices there are only $n - u$ potential new neighbors for a vertex w. Thus its number of new neighbors has Binomial Distribution with parameters $n - u, \lambda/n$. If u is large the estimation by the Poisson P_λ is no longer valid. These technical problems are all resolvable and it turns out that their asymptotic effect is nil. Identifying new neighbor with child gives the key correspondence: The limiting distribution of the r-neighborhood of v is asymptotically that of the (r, λ) Poisson tree of Eve.

What are the probabilities for first order sentences on the (r, λ) Poisson tree T? We examine the (r, s)-value as defined in section 3.3.1. For $r = 1$ these values are that Eve has $0, 1, \ldots, s$ or M (meaning more than s) children. We set $p_i = e^{-\lambda}\lambda^i/i!$ for $0 \le i \le s$ and $p_M = 1 - \sum_{i \le s} p_i$, so that p_α is the probability of having value α. What about $r = 2$. First Eve has P_λ children and then each of the children gets $(1, s)$ value α with probability p_α. Here a special property of the Poisson distribution comes into play. This is equivalent to Eve having $P_{p_\alpha \lambda}$ children with $(1, s)$ value α, *independently* for each α. [For example, if Alice has P_6 children and each child is sanguine with probability 2/3 and morose with probability 1/3 then this is equivalent to Alice having P_4 sanguine children and, independently, P_2 morose children.] Consider a $(2, s)$ value given by a function $u(\alpha)$. For each α the probability of having $i = u(\alpha)$ children with $(1, s)$ value α is then $e^{-\mu}\mu^i/i!$ (or one minus the sum of such terms when $i = M$) where $\mu = p_\alpha \alpha$. The probability that the $(2, s)$ value is given by the function u is the product of these terms (by the independence) over all possible α.

Example: We take $s = 0$ so that the $(1, 0)$ values are no children and some children with probabilities $e^{-\lambda}$ and $1 - e^{-\lambda}$. Consider the $(2, 0)$ value of having no children with no children and some children with some children. No children with no children has probability $\exp[-\lambda e^{-\lambda}]$ and some children with some children has probability $1 - \exp[-\lambda(1 - e^{-\lambda})]$ and the $(2, 0)$ value has probability their product.

In general suppose the $(r - 1, s)$ values have probabilities p_β. The probability of having i children with $(r - 1, s)$ value β is then $e^{-\mu}\mu^i/i!$ ($i \ne M$, and one minus that sum for $i = M$) where $\mu = \lambda p_\beta$. Any particular (r, s) value is the product of those terms over β.

Example: The probability of having no children with no children was $\exp[-\lambda e^{-\lambda}]$ and so the probability of having no children with no children with no children is $\exp\big[-\lambda \exp[-\lambda e^{-\lambda}]\big]$.

With the exact description of the probabilities for (r, s) values is messy, their type, as functions of λ can be easily described. They are generated from λ by products, sums, difference, constant multiplication and, critically, base e exponentiation. Unlike the situation of section 3.6.4 when $p = \frac{\ln n}{n} + \frac{c}{n}$, here the base e exponentiation may be iterated an arbitrary number (namely r) times. We note that these are all continuous functions for λ positive and that $\lambda = 1$ plays no special role.

It is helpful here to think of $G(n, \lambda/n)$ and the first order language with the addition of a constant symbol v. The r-neighborhood of v has the asymptotic distribution of (r, λ) Poisson tree. First order sentences in the language of rooted trees could then be translated into first order sentences in our graph language. For example, our favorite "Eve has no children that have no children that have no children" becomes

$$\neg\exists_w \big(w \sim v \wedge \neg\exists_x \big[x \sim w \wedge x \ne v \wedge \neg\exists_y [y \sim x \wedge y \ne w]\big]\big)$$

One can translate any (r, s) values into first order sentences – the formal proof being an induction on r. These first order sentences now have limiting probabilities which are functions of λ as described above.

Our first order language does *not* actually contain constant symbols. Basically, the vertices in small cycles are special and will be treated like constant symbols since these can be distinguished through a first order lens. Let $k \geq 3, r, \lambda$ be given and define an (r, λ) Poisson k-cycle as follows. Begin with a k-cycle. From each vertex of the k-cycle generate independently disjoint (r, λ) Poisson trees.

Consider such a unicyclic graph. Label the vertices of the cycle $1, \ldots, k$ in order. For each such vertex i let β_i denote the (r, s)-value of the tree coming off that vertex. We call $\boldsymbol{\beta} = (\beta_1, \ldots, \beta_k)$ the (r, s)-value of the unicyclic graph with the proviso that it should be independent of the particular labelling – for example, with $k = 4$, $(\beta_1, \beta_2, \beta_3, \beta_4)$, $(\beta_2, \beta_3, \beta_4, \beta_1)$ and $(\beta_4, \beta_3, \beta_2, \beta_1)$ are all regarded the same. The probability that an (r, λ) Poisson k-cycle has a particular (r, s) value is then simply the product of the probabilities for the k (r, λ) Poisson trees times a constant for the number of placements of the k-tuple. (For example, when $\beta_1, \beta_2, \beta_3, \beta_4$ are distinct there are eight placements while when $\beta_1 = \beta_2 = \beta_3 \neq \beta_4$ there are four placements.) As before, this probability considered as a function of λ can be generated from λ through addition, subtraction, multiplication, constant multiplication and base e exponentiation.

3.7.2 Almost Completing the Almost Sure Theory

The almost sure theory T for $G(n, \lambda/n)$ (any positive λ) is generated by two axiom schema:

1. For every k: There do not exist k vertices with (at least) $k + 1$ edges.
2. For every finite tree T (including the "one point tree") and every r: There exist (at least) r components T

The countable models can now be completely described. Every component is either a tree or unicyclic. Every finite tree occurs infinitely often as a component. Every infinite tree may or may not occur as a component. Every unicyclic graph may or may not occur as a component.

In the range $p = \lambda/n$ the number of k-cycles, for any fixed $k \geq 3$, can be arbitrary. For example, there is a positive limiting probability that at $p = 2/n$ there are precisely five hexagons. Further, the local neighborhoods of those k-cycles can be arbitrary. There is a positive limiting probability that there are precisely seven triangles whose vertices have $4, 17, 5$ neighbors.

In our previous cases we have added sentences σ_i to a theory T so that each $T + \sigma_i$ was complete. Here we are a bit less stringent. Let s be arbitrary.

Definition 3.2 *A sentence σ_f is an s-*completion *of T if for any first order A of quantifier depth at most s either A or $\neg A$ is in $T + \sigma_f$.*

Applying Theorem 2.3.1, σ_f is an s-completion of T if and only if Duplicator wins $\text{EHR}(G_1, G_2; s)$ for any two models G_1, G_2 of $T + \sigma_f$.

Set $d = \frac{3^s-1}{2}$. Let Γ be a list containing for each $3 \le k \le 2d+1$ the possible (d,s)-values $\boldsymbol{\beta} = (\beta_1, \ldots, \beta_k)$ of k-cycles. Let f be a function from Γ to the nonnegative integers. Let σ_f be the first order statement that for each such $k, \boldsymbol{\beta}$ there are precisely $f(\boldsymbol{\beta})$ k-cycles with that (d,s)-value.

We claim that these σ_f are s-completions of T. For suppose G_1, G_2 both satisfy T and the same σ_f. Let S_1, S_2 denote the vertices of G_1, G_2 respectively in cycles of length at most d. These may be matched up cycle for cycle so that the first condition of Theorem 2.6.7 is satisfied. Suppose $x \in G_1$ with $\rho(x, s_1) > 2d' + 1$ for all $s_1 \in S_1$. Then the d'-neighborhood of x is a rooted tree which has some (d,s) value α. But G_2 has infinitely many copies of every finite tree as a component. Thus for all $y_1, \ldots, y_{s-1} \in G_2$ there will exist $y \in G_2$ whose d'-neighborhood is s-similar to that of x where y lies in a different component from all of S_2 and all $y_1, \ldots, y_{s-1}$. This, and the symmetric version, give the other conditions for Theorem 2.6.7. Hence $G_1 \equiv_k G_2$, giving the claim.

There is one further reduction. If any $f(\boldsymbol{\beta}) \ge s$ we may simply call it M, for many. Spoiler will not be able to take advantage of the difference of the numbers in an s-move game. The truncated information is then a function $f\colon \Gamma \to \{0, \ldots, s-1, M\}$. But conversely the information given by f is itself expressible by a first order sentence. Hence the possible limiting probabilities of first order sentences are precisely the possible limiting probabilities for the truncated information L'.

The above limiting probabilities can be easily described. Fix A with quantifier depth s. Set $p = \lambda/n$ and $d = \frac{3^s-1}{2}$. For each $3 \le k \le 2d$ the expected number of k-cycles is $\lambda^k/2k$. The asymptotic probability $p(\boldsymbol{\beta})$ that a k-cycle has (examining its d-neighborhood) a given type $\boldsymbol{\beta} = (\beta_1, \ldots, \beta_k)$ is given by the probability that the (r, λ) Poisson k-cycle has that type. Let $X_{\boldsymbol{\beta}}$ denote the number of k-cycles of type $\boldsymbol{\beta}$. Then $X_{\boldsymbol{\beta}}$ has Poisson distribution with mean $p(\boldsymbol{\beta})\lambda^k/2k$. Further, the $X_{\boldsymbol{\beta}}$ are mutually independent over all choices of k and $\boldsymbol{\beta}$. With s fixed there are only finitely many such choices. Any limiting probability is then expressible as the finite sum of finite products of terms of the form $\Pr[X_{\boldsymbol{\beta}} = i]$ (where $i \le s$) or $\Pr[X_{\boldsymbol{\beta}} \ge s]$. We have already seen that the $p(\boldsymbol{\beta})$ can be expressed as a function of λ using λ, constants, addition, subtraction, multiplication and base e exponentiation. Thus the same holds for the means $p(\boldsymbol{\beta})\lambda^k/2k$ of the $X_{\boldsymbol{\beta}}$ Our conclusion was first shown in [14]:

Theorem 3.7.1 (Lynch). *For every first order A*

$$\lim_{n\to\infty} \Pr\bigl[G(n, \lambda/n) \models A\bigr]$$

exists and can be expressed as a function of λ using λ, constants, addition, subtraction, multiplication and base e exponentiation.

Example: Call a triangle v_1, v_2, v_3 bushy if every neighbor w of any of the v_i (other than another of the v_j) itself has at least one neighbor other than v_i itself. In other words: looking at the rooted trees from each v_i given by deleting the other v_j – each of the three rooted trees has the property that the root has no children that have no children. Consider the event A: "There are no bushy triangles" The expected number of triangles is $\lambda^3/6$. Each vertex of such a triangle generates a rooted tree with no children having no children with probability $\exp[-\lambda e^{-\lambda}]$. The probability that the triangle is bushy is then $\exp[-3\lambda e^{-\lambda}]$. The expected number of bushy triangles is then $(\lambda^3/6)\exp[-3\lambda e^{-\lambda}]$. This number has a Poisson distribution and so $\Pr[A]$ has limiting probability $\exp\left[-(\lambda^3/6)\exp[-3\lambda e^{-\lambda}]\right]$.

4. The Combinatorics of Rooted Graphs

Formally a rooted graph is simply a graph H with a designated subset R of vertices, called the roots. We allow the degenerate case $R = \emptyset$ but not $R = V(H)$. The rooted graph is denoted by the pair (R, H). For our purposes rooted graphs are a convenient notation for discussing the extension statements $\mathrm{Ext}(R, H)$ defined in Section 1.3 and our definitions will be meant to reflect the almost always character of random graphs. But we also feel that the theory of rooted graphs is a fascinating combinatorial structure of interest in its own right.

4.1 Sparse, Dense, Rigid, Safe

A rooted graph (R, H) has two main parameters: the number of vertices that are not roots, generally denoted v, and the number of edges, not counting edges with both vertices being roots, generally denoted e. We call (v, e) the *type* of (R, H). A third parameter is the number of roots, generally denoted r. Perhaps surprisingly, this value plays little role in the theory.

Definition 4.1 *The* type *of a rooted graph (R, H) is the pair (v, e) where v is the number of vertices that are not roots and e is the number of edges of H which do not have both vertices in R.*

Throughout this chapter α shall be a fixed positive *irrational* real number. The definitions of sparse, dense, etc., shall be all relative to this α and are meant to be meaningful in studying $G(n, p)$ with $p = n^{-\alpha}$.

Definition 4.2 *Let (R, H) have type (v, e). If $v - e\alpha$ is positive we call (R, H)* sparse. *If $v - e\alpha$ is negative we call (R, H)* dense.

Because α is irrational and v, e are, of course, integers there is a *strict dichotomy* – all rooted graphs are either sparse or dense. In $G(n, p)$ fixing roots $\mathbf{x}$ the expected number of (R, H) extensions $\mathbf{y}$ is $\sim cn^v p^e = cn^{v-e\alpha}$. For dense extensions this is a negative power of n so that almost surely a given $\mathbf{x}$ will not have such $\mathbf{y}$. For sparse extensions we get a positive power of n. But expectation going to infinity does not, a priori, imply that an event occurs almost surely and indeed this will not always be the case.

Definition 4.3 *Let* $R \subset S \subseteq V(H)$. *We call* $(R, H|_S)$ *a* subextension *of* (R, H). *Let* $R \subseteq S \subset V(H)$. *We call* (S, H) *a* nailextension *of* (R, H).

The idea is that we are "nailing down" the further vertices $S - R$ and making them roots. Note that a rooted graph is both a subextension and a nailextension of itself. We call a subextension or nailextension *nontrivial* if it is not (R, H) itself.

We allow some latitude for notational convenience. When the rooted graph (R, H) is understood we shall sometimes write subextensions in the form (R, S), by which we mean $(R, H|_S)$ or (R, H_1) where $H_1 = H|_S$. Nailextensions (S, H) may be written (H_1, H) where $H_1 = H|_S$. This should not cause confusion as a rooted graph is always, formally, a pair consisting of a set followed by a graph. We may have chains $R = R_0 \subset R_1 \subset \ldots \subset R_{t-1} \subset R_t = V(H)$. We may then naturally decompose (R, H) into subextensions, formally $\left(R_i, H|_{R_{i+1}}\right)$, which we may write as either (R_i, R_{i+1}) or (H_i, H_{i+1}) where $H_i = H|_{R_i}$. We also have a sequence of nailextensions (R_i, H) which we may also write as (H_i, H).

Definition 4.4 *An extension is called* rigid *if all of its nailextensions are dense. An extension is called* safe *if all of its subextensions are sparse.*

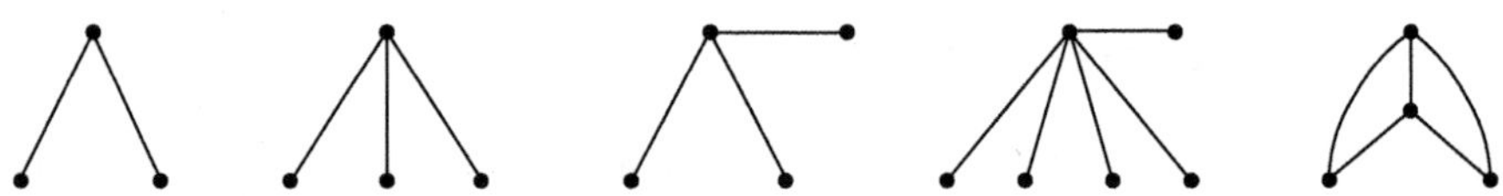

Fig. 4.1. Rooted graphs with roots at bottom level

For examples we shall set $\alpha = \pi/7 = 0.4487\cdots$. This value has no special properties and so provides good generic examples of our properties. In the figure above the rooted graphs have types $(1,2)$, $(1,3)$, $(2,3)$, $(2,5)$,$(2,5)$ and so are sparse, dense, sparse, dense and dense respectively. The first two are safe and rigid, indeed when there is only $v = 1$ nonroot there are no nontrivial sub- or nailextensions. The third is sparse but is not safe as it has the first as a subextension. The fourth is dense but not rigid as if the vertex adjacent to the four roots is itself made a root (nailed down) the resulting nailextension has type $(1,1)$ and is sparse. The fifth is rigid as nailing down a nonroot gives a nailextension of type $(1,3)$ which is dense. We shall see that safe and rigid are the appropriate meanings for completely sparse and completely dense. Observe for now that $\mathrm{Ext}(R, H)$ does *not* hold almost surely in the third case, roots x_1, x_2, x_3 will not have an extension y_1, y_2 since they will not have a common neighbor y_1.

Let $R \subset S \subset V(H)$. Let (R, H) have type (v, e). Let the types of $(R, H|_S)$ and (S, H) be (v_1, e_1) and (v_2, e_2) respectively. We then note that

$$e = e_1 + e_2 \text{ and } v = v_1 + v_2$$

As $v - e\alpha = (v_1 - \alpha e_1) + (v_2 - \alpha e_2)$ we have

Property 4.1.1 *The dense extension of a dense extension is dense.*

Property 4.1.2 *The sparse extension of a sparse extension is sparse.*

We further claim

Property 4.1.3 *Any sparse extension has a safe nailextension.*

Proof: Let (R, H) be sparse. If it itself is not safe then some subextension (R, S) will be not sparse, and hence dense. Pick a maximal such S, which cannot be $V(H)$ as (R, H) is assumed sparse. Then (S, H) is safe. For if it were not there would be a dense subextension (S, T) and then (R, T) would be dense, contradicting the maximality of S.

Property 4.1.4 *Any dense extension has a rigid subextension.*

Proof: Let (R, H) be dense. If it itself is not rigid then some nailextension (S, H) will not be dense, and hence sparse. Pick a minimal such S, which cannot be R as (R, H) is assumed dense. Then (R, S) is rigid. For if it were not there would be a sparse nailextension (T, S) and then (T, H) would be sparse, contradicting the minimality of S.

The above results demonstrate an intriguing duality in these concepts. Here are two useful tautological statements:

Property 4.1.5 *All subextensions of a safe extension are safe.*

Property 4.1.6 *All nailextensions of a rigid extension are rigid.*

We further note

Property 4.1.7 *An extension which is not safe has a rigid subextension.*

Property 4.1.8 *An extension which is not rigid has a safe nailextension.*

Proof: If (R, H) is not safe it has a dense subextension (R, H') which in turn has a rigid subextension (R, H^*). If (R, H) is not rigid it has a sparse nailextension (R', H) which in turn has a safe nailextension (R^*, H).

Now let (R, H) be a rooted graph in some larger graph and let X be a set of vertices disjoint from $V(H)$. By $(R \cup X, H \cup X)$ we mean the graph on $V(H) \cup X$, with $R \cup X$ as the roots. That is, we add new roots to (R, H) and (possibly) new edges.

Property 4.1.9 *If (R, H) is sparse and $X \cap V(H) = \emptyset$ and there are no edges between X and $V(H) - R$ then $(R \cup X, H \cup X)$ is sparse.*

The type (v, e) of the extension has remained the same.

Property 4.1.10 *If (R, H) is safe and $X \cap V(H) = \emptyset$ and there are no edges between X and $V(H) - R$ then $(R \cup X, H \cup X)$ is safe.*

Any subextension of $(R \cup X, H \cup X)$ may be written $(R \cup X, H' \cup X)$ where (R, H') is a subextension of (R, H). As (R, H) is safe, (R, H') is sparse so $(R \cup X, H' \cup X)$ is sparse by Property 4.1.9.

Property 4.1.11 *If (R, H) is dense and $X \cap V(H) = \emptyset$ then $(R \cup X, H \cup X)$ is dense.*

The number of nonroots v has remained the same and all e edges counted in (R, H) are still counted in $(R \cup X, H \cup X)$.

Property 4.1.12 *If (R, H) is rigid and $R \cup X \neq V(H) \cup X$ then $(R \cup X, H \cup X)$ is rigid.*

As $x \in X \cap R$ do not affect the conclusion let us assume for convenience that $X \cap R = \emptyset$. Write $X = Y \cup Z$ with $Y = V(H) \cap X$, $Z = X - Y$. Any nailextension of $(R \cup X, H \cup X)$ can be expressed as $(R \cup X \cup W, H \cup X)$ where $W \subset V(H) - R - Y$. Then $(R \cup Y \cup W, H)$, being a nailextension of (R, H), is dense. Apply Property 4.1.11, adding vertices Z. Then $(R \cup X \cup W, X \cup X)$ is dense.

Property 4.1.13 *A rigid extension of a rigid extension is rigid.*

Let (R, H) and (H, H_1) be rigid and let $R \subseteq S \subset V(H_1)$. Then $\big(V(H) \cap S, H\big)$ is dense so, adding new roots $S - V(H)$, $(S, H \cup S)$ is dense. As (H, H_1) is rigid $(H \cup S, H_1)$ is dense and hence by Property 4.1.1 (S, H_1) is dense. As S was arbitrary (R, H_1) is a rigid extension.

Property 4.1.14 *A safe extension of a safe extension is safe.*

Let (R, H) and (H, H_1) be safe and let $R \subset S \subseteq V(H_1)$. Then $\big(R, V(H) \cap S\big)$ is a subextension of a safe extension and hence sparse. $\big(V(H), V(H) \cup S\big)$ is also a subextension of a safe extension and hence sparse. Remove the roots $V(H) - S$ from the extension $\big(V(H), V(H) \cup S\big)$. The number of nonroots remains the same and the number of edges decreases or remains the same. Hence the remaining extension $\big(V(H) \cap S, V(H) \cup S\big)$ is also sparse. (R, S) is now sparse by Property 4.1.2. As S was arbitrary (R, H_1) is a safe extension.

Property 4.1.15 *If (R, H) is neither safe nor rigid there exists S with $R \subset S \subset V(H)$ such that (R, S) is rigid and (S, H) is safe.*

Proof: By Property 4.1.7 there is a rigid extension. Let S be maximal with (R, S) rigid. Were (S, H) to have any rigid subextensions (S, T) then by Property 4.1.13 (R, T) would be rigid, contradicting the maximality. Thus (S, H) has no rigid subextensions so by Property 4.1.4 it has no dense subextensions so all subextensions are sparse. Thus (S, H) is safe.

Definition 4.5 *We call (R, H)* minimally safe *if it is safe and there is no S with $R \subset S \subset V(H)$ and both (R, S) and (S, H) safe.*

Property 4.1.16 *If (R, H) is minimally safe and $R \subset S \subset H$ then (S, H) is dense.*

Proof: From Property 4.1.5 (R, S) is safe, hence (S, H) cannot be safe. If (S, H) is rigid it is dense. Otherwise we apply Property 4.1.15 to (S, H) to find T with $S \subset T \subset V(H)$ with (S, T) rigid and (T, H) safe. Again by Property 4.1.5 (R, T) is safe so that T contradicts the assumption that (R, H) is minimally safe.

4.2 The t-Closure

Recall Definition 1.5 that $\mathbf{y}$ is an (R, H)-extension of $\mathbf{x}$. This leads us to a central definition that becomes the key to understanding first order properties on random graphs.

Definition 4.6 *Let α be a fixed positive irrational real and t a fixed positive integer. Let G be a finite graph and U a subset of its vertices. The t-*closure *of U, denoted $\text{cl}_t(U)$ is the minimal set of vertices X satisfying*
- *$U \subseteq X$*
- *Let (R, H) be a rigid (with respect to the given α) rooted graph with r roots and $v \leq t$ nonroot vertices . Let $\mathbf{x} = (x_1, \ldots, x_r)$ be an r-tuple of distinct vertices of X. If $\mathbf{y} = (y_1, \ldots, y_v)$ is an (R, H) extension of $\mathbf{x}$ then all $y_1, \ldots, y_v \in X$.*

As $\text{cl}_t(U)$ can be found by a finite sequence of rigid extensions we have from Property 4.1.13 that $\text{cl}_t(U)$ is a rigid extension of U, including the important possibility that it is simply equal to U itself.

Notice that given α and t there are only a finite number of (R, H) extensions to look at. Suppose, for example, $\alpha = \pi/7$ and $t = 1$. The only way y can be a rigid extension of x is if y consists of one vertex adjacent to at least three of the vertices of x. So $\text{cl}_1(U)$ may be described by: Start with U and extend for as long as possible by vertices adjacent to at least three vertices in your set.

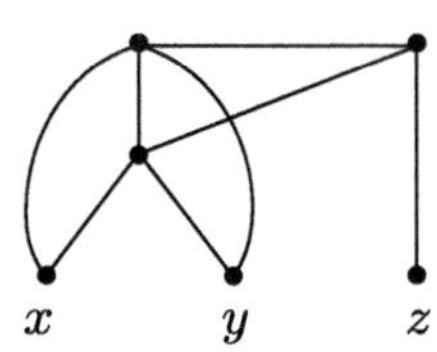

Fig. 4.2. A possible $\text{cl}_2(x, y, z)$

We may view $y \in \text{cl}_t(U)$ as meaning that y is special with respect to U. The gradation of specialness by the parameter t is critical to the theory.

Let G, G' be graphs with r-tuples $\mathbf{x} = (x_1, \ldots, x_r)$, $\mathbf{y} = (y_1, \ldots, y_r)$ respectively.

Definition 4.7 *We say* $\mathrm{cl}_t(\mathbf{x})$ *is isomorphic to* $\mathrm{cl}_t(\mathbf{y})$, *and write* $\mathrm{cl}_t(\mathbf{x}) \cong \mathrm{cl}_t(\mathbf{y})$ *if there is a graph isomorphism from* $\mathrm{cl}_t(\{x_1, \ldots, x_r\})$ *to* $\mathrm{cl}_t(\{y_1, \ldots, y_r\})$ *that sends* x_i *to* y_i *for* $1 \le i \le r$. *The* t-type *of* $\mathbf{x}$ *is its equivalence class under the above isomorphism.*

Definition 4.8 *Let* H *be a graph with designated distinct vertices* $a_1, \ldots, a_r$. *We write* $\mathrm{cl}_t(\mathbf{x}) \cong H$ *if there is a graph isomorphism from* $\mathrm{cl}_t(\{x_1, \ldots, x_r\})$ *to* H *which sends* x_i *to* a_i *for* $1 \le i \le r$.

The property $\mathrm{cl}_t(\mathbf{x}) \cong H$ is described by a first order predicate over the x_i: that first there exist y_j so that the x_i, y_j form precisely the finite graph H and second that there do not exist z_k distinct from the x's and y's forming any one of the finitely many types of rigid extensions of at most t vertices over the x's and y's.

The t-type can often be described in words. For $\alpha = \pi/7$, here are some possible 1-types of a triple x_1, x_2, x_3:

1. x_1, x_2, x_3 have three common neighbors y_1, y_2, y_3 which are mutually nonadjacent and no three of the above six points have any further common neighbor. There is precisely one adjacency amongst the roots.
2. x_1, x_2, x_3 have a common neighbor y_1 and there is a common neighbor y_2 of x_1, x_2, y_1 and there are no further adjacencies amongst these vertices nor further common neighbors amongst three of these points.
3. x_1, x_2 are adjacent, nonadjacent to x_3 and the three have no common neighbor.
4. x_1, x_2, x_3 are nonadjacent and have no common neighbor

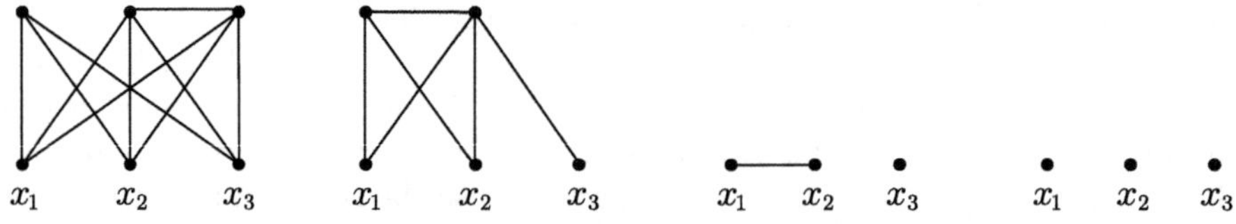

Fig. 4.3. Possible $\mathrm{cl}_1(x_1, x_2, x_3)$

The above figure illustrates these possibilities with pictures, where there must be the understanding that there are no further vertices nor edges than what is in the picture. The final possibility can be generalized – we say the t-closure of $x = (x_1, \ldots, x_r)$ is *trivial* if the x_i, x_j are all nonadjacent and the t-closure consists only of $x_1, \ldots, x_r$.

4.3 The Finite Closure Theorem

While the t-closure, as given by Definition 4.6, is defined for any graph G it has been designed for use on the random graph $G \sim G(n, p)$ with $p = n^{-\alpha}$.

Theorem 4.3.1. *Fix α, t, r, let $G \sim G(n,p)$ with $p = n^{-\alpha}$ and let $x_1, \ldots, x_r$ be selected randomly. Then almost surely the t-closure of $\mathbf{x} = (x_1, \ldots, x_r)$ is trivial.*

Proof: As $p = o(1)$ and r is fixed almost surely there are no adjacencies between the x_i. Let (R, H) be a rigid extension of type (v, e). The expected number of (R, H) extensions $\mathbf{y}$ of $\mathbf{x}$ is $\sim cn^v p^e$ which is a negative power of n (by denseness) and hence $o(1)$. Almost surely there are no such $\mathbf{y}$. But there are only a finite number of possible (R, H) extensions (as $V(H) \leq r+t$, surely at most 2 to the power $\binom{r+t}{2}$) so almost surely for every such (R, H) there is no extension $\mathbf{y}$.

While most $\mathbf{x}$ have trivial t-closure it is most certainly not the case that all $\mathbf{x}$ will have trivial t-closure. Taking $\alpha = \pi/7$ and $t = 1$ the average vertex y has $\sim n^{1-\alpha}$ neighbors so it certainly has three neighbors x_1, x_2, x_3. Then $\mathbf{x} = (x_1, x_2, x_3)$ has a nontrivial 1-closure, as it includes the vertex y.

Theorem 4.3.2 (Finite Closure Theorem). *Given α, t, r there is a constant K such that in $G \sim G(n,p)$ with $p = n^{-\alpha}$ almost surely every r-tuple $\mathbf{x}$ has t-closure of size less than $r + K$.*

We first give an illustrative example. With $\alpha = \pi/7$, $r = 3$, $t = 1$ are there three vertices with 100 common neighbors? No, almost surely that will not occur anywhere in the graph. For if it did we would have a graph on $3 + 100$ vertices (the 3 roots plus the 100 nonroots) and $3(100)$ edges (each nonroot having three edges to the roots) and the expected number of such structures is $O(n^{103}p^{300})$. The power of n here is $3 + 100(1 - 3\alpha) = 3 + 100(-0.34\cdots)$ is most certainly negative and so almost surely no such structure exists.

Proof: Set

$$\beta = \min \frac{e\alpha - v}{v}$$

the minimum over pairs of integers v, e with $v \leq t$ and $v - e\alpha < 0$. There are only finitely many v to consider and for each only one e (the smallest) to consider so the minimum does exist and is positive. We show Theorem 4.3.2 for

$$K = \left\lceil \frac{r}{\beta} \right\rceil$$

If there were a $\mathrm{cl}_t(R)$ of size at least K, with $|R| = r$, we would have a sequence $R = S_0 \subset S_1 \ldots \subset S_u \subseteq \mathrm{cl}_t(R)$ where each (S_i, S_{i+1}) was rigid of type (v_i, e_i) with $v_i \leq t$ and $K \leq \sum_{i=0}^{u-1} v_i \leq K + t$. The graph on S_u would then have $V = r + \sum v_i$ vertices and at least $E = \sum e_i$ edges. The expected number of such graphs is $O(n^{V - \alpha E})$. But

$$V - \alpha E = r + \sum (v_i - \alpha e_i) \leq r - \sum_i \beta v_i \leq r - K\beta$$

and we have chosen K so that this is negative. Thus almost surely this sequence never occurs but as there are only a bounded number of sequences to consider almost surely there is no such sequence so that all $\mathrm{cl}_t(R)$ have size less than K.

Effectively each rigid extension above has a cost (in calculating $n^v p^e$) which is at least β per vertex gained. We begin with capital r, the r degrees of freedom of the roots, but cannot continue paying out the costs beyond a certain point.

The upper bound K given above depends in a striking way upon α. This is not apparent for $\alpha = \pi/7$ (as, indeed, we selected this to be a typical α) but becomes clear when we set $\alpha = \frac{1}{2} + \epsilon$ and consider ϵ small and positive. Let $r = 2$ and $t = 1$ so that the only rigid extension is adding a vertex adjacent to two vertices already in the set. Then $\beta = 2\alpha - 1 = 2\epsilon$ so that K in the above theorem is taken to be $\lceil \epsilon^{-1} \rceil$. Perhaps surprisingly, this result is best possible . The complete bipartite graph $K_{2,t}$ is known from the classical work or Erdős and Rényi [5] to almost always exist at $p = n^{-\alpha}$ if $n^{t+2}p^{2t}$ has a positive power of n which occurs for $t\epsilon < 1$, thus for $t = \lfloor \epsilon^{-1} \rfloor$. When $K_{2,t}$ exists the two bottom points have t common neighbors and therefore a 1-closure of at least t more points. Similar (though not quite as precise) results occur for other r, t.

If we think of r, t fixed and $K = K(\alpha)$ then the function K has intriguing properties. At every rational v/e with $v \le t$ there is a one-sided pole. As α approaches v/e from above $K(\alpha)$ approaches infinity in a hyperbolic fashion. That is, $K(v/e + \epsilon) \sim c/\epsilon$ with c a constant depending on v, e, r. But as α approaches v/e from below there is no problem and $K(\alpha)$ approaches a finite limit. The nonuniformity of $K(\alpha)$ leads to interesting complications when we look dynamically at the behavior of a fixed sentence A at $G(n, n^{-\alpha})$ as α varies. We return to this topic in Section 9.2.2.

For completeness it is useful to include two special cases of the t-closure and t-type. First $t = 0$. The 0-closure of a set X is simply X itself. Two k-tuples $(x_1, \ldots, x_k) \in G_1$, $(x'_1, \ldots, x'_k) \in G_2$ have the same 0-type if they are the same induced subgraphs: more precisely that $x_i \sim x_j$ if and only if $x'_i \sim x'_j$. Duplicator wins the Ehrenfeucht game $\textsc{Ehr}(G_1, G_2; k)$ if at the end of the game the k-tuples of vertices selected (in order of selection) in each graph have the same 0-type. Second, $k = 0$ points – the t-closure of the empty set. A rooted graph $(\emptyset, H)$ is dense if and only if $v - e\alpha < 0$ where H has v vertices and e edges. For $(\emptyset, H)$ to be rigid this is a necessary condition. For $\mathrm{cl}_t(\emptyset)$ to be nonempty there must be a rigid extension $(\emptyset, H)$ where H has $v \le t$ vertices and e edges and $v - e\alpha < 0$.

Theorem 4.3.3. *Almost surely* $\mathrm{cl}_t(\emptyset) = \emptyset$.

Proof: For each H with v vertices and e edges as above the expected number of copies of H is $O(n^v p^e) = O(n^{v - \alpha e})$ which is $o(1)$. Each H almost surely doesn't appear so almost surely none of the finitely many possible H appear.

This gives us the beginning and end of the Duplicator look-ahead strategy described in Section 6.1. At the start of the game (before any points have been played) the t-closures (t will be selected suitably large) of $G_1 \sim G(n, n^{-\alpha})$ and $G_2 \sim G(m, m^{-\alpha})$ are almost surely the same as they are both null. At the end of the game Duplicator wants the 0-closures to be the same. Analogously to the examples of Section 2.4 she will keep the u-closures the same where u will decrease (cleverly) as the game progresses.

5. The Janson Inequality

The Janson Inequality is a relatively recent result that has proven to be very useful in the study of random graphs. We begin with a general (though hardly the most general) formulation. An important and instructive example of its application is given after the statement of the Inequality.

Let Ω be a finite set. Let $R \subseteq \Omega$ be a random subset of Ω with $\Pr[e \in R] = p_e \in [0,1]$ for each $e \in \Omega$ where, critically, the events $e \in R$ are mutually independent over the $e \in \Omega$. That is, each $e \in \Omega$ flips its own coin to decide if it is in R. Let I be a finite index set and $A_i, i \in I$ be subsets of Ω. Let B_i be the event $R \supseteq A_i$, that all the $e \in A_i$ flipped their coins "heads". The Janson Inequality gives bounds on the event

$$B := \bigwedge_{i \in I} \overline{B_i}$$

For all $i \in I$ set $p_i = \Pr[B_i]$ and set

$$M := \prod_{i \in I} (1 - p_i)$$

Informally, M is what $\Pr[B]$ would be if the events B_i were mutually independent. This occurs if the sets A_i are mutually disjoint which we do *not* assume. For future reference we set

$$\mu = \sum_{i \in I} \Pr[B_i]$$

We let δ be such that

$$\Pr[B_i] \le \delta$$

for all $i \in I$. We write $i \sim j$ for $i, j \in I$ if they are unequal and $A_i \cap A_j \neq \emptyset$. A key role is played by

$$\Delta = \sum_{i \sim j} \Pr[B_i \wedge B_j],$$

the sum over ordered pairs i, j. The relation $\sim$ is meant to capture dependency among the B_i and Δ is meant to give a quantitative measure of the dependency.

Theorem 5.0.4 (Janson's Inequality). *Under the above assumptions*

$$M \le \Pr[B] \le Me^{\Delta/(2(1-\delta))}$$

As discussed in Section 0.3, where full references are given, Janson's Inequality is a notable exception to our policy of keeping this work self-contained. We content ourselves here with a statement of the result and a description of various applications.

For application to random graphs we set $\Omega = \{\{i,j\}, 1 \le i < j \le n\}$, the potential edges of $G(n,p)$. Each $e \in \Omega$ has $p_e = p$ so that R is the (random) edge set of $G(n,p)$. Now fix a vertex x of $G(n,p)$. Let the index set I range over all pairs $\{y,z\} \subset \{1,\ldots,n\} - \{x\}$. For each such yz (ignoring set brackets for convenience) let $A_{yz} = \{\{x,y\},\{x,z\},\{y,z\}\}$. Then B_{yz} is the event that xyz is a triangle and B is the event that x lies in no triangle. Now set $p = n^{-2/3+\epsilon}$ where $0 < \epsilon < \frac{1}{15}$ is fixed. We may take $\delta = \Pr[A_{yz}] = p^3$. We calculate

$$M = (1-p^3)^{\binom{n-1}{2}} \sim e^{-n^2p^3/2} \sim e^{-n^{3\epsilon}/2}$$

The relation $\sim$ holds when pairs overlap, so they must be of the form $\{y,z\} \sim \{y,z'\}$. There are $O(n^3)$ such pairs. For each $B_{y,z} \wedge B_{y,z'}$ requires the five edges xy, xz, yz, xz', yz' to be in $G(n,p)$ and so has probability p^5. Thus $\Delta = O(n^3p^5)$. Our range for ϵ was designed so that $\Delta = o(1)$.

We shall be particularly interested in situations when $\delta = o(1)$ and, more importantly, $\Delta = o(1)$. In that case $e^{\Delta/(2(1-\delta))} \sim 1$ and so we have an *asymptotic formula* $\Pr[B] \sim M$. As all $p_i = o(1)$ all $1 - p_i = e^{-p_i(1+o(1))}$ so that $M = e^{-\mu(1+o(1))}$. Indeed, in many cases calculation gives M asymptotic to $e^{-\mu}$. We do have $\delta = o(1)$ and $\Delta = o(1)$ in our graph example so that the probability x does not lie in a triangle is asymptotic to M. With ϵ fixed $M = o(n^{-c})$ for any constant c. In particular $M = o(n^{-1})$. Now consider the first order event E that *every* vertex lies in a triangle. Each vertex has probability $M = o(n^{-1})$ of not lying in a triangle and there are "only" n vertices. Thus $\Pr[\neg E] \le n \cdot o(n^{-1}) = o(1)$ and so E holds almost surely.

5.1 Extension Statements

Clearly, the property "Every vertex lies in a triangle" is just one example of the extension statements $\mathrm{Ext}(R,H)$ as defined in Section 1.3. Can we generalize the application of Janson's Inequality to other $\mathrm{Ext}(R,H)$? The answer is yes *if* we add on some important restrictions.

From Definition 4.1 we have the type (v,e) of a rooted graph (R,H) and of a subextension $(R,H|_S)$.

Definition 5.1 *A rooted graph (R,H) is* balanced *if the maximal ratio e'/v' amongst all subextensions $(R,H|_S)$ (with (v',e') the type) is achieved for (R,H) itself and* strictly balanced *if that ratio was is achieved only for (R,H).*

When $R = \emptyset$ this reduces to Definition 1.2.

Theorem 5.1.1. *Let (R, H) be strictly balanced with type (v, e) and set $a = v/e$. Let $p = n^{-a+\epsilon}$ with ϵ positive. Then* $\mathrm{Ext}(R, H)$ *holds almost surely in $G(n, p)$.*

Proof: Fix a choice of roots $\mathbf{x} = (x_1, \ldots, x_r)$, the x_i distinct vertices of $G \sim G(n, p)$. For $\mathbf{y} = (y_1, \ldots, y_v)$ let $B_\mathbf{y}$ be the event that $\mathbf{y}$ forms an (R, H) extension over $\mathbf{x}$. Set $B = \bigwedge \overline{B_\mathbf{y}}$, the event that there is no (R, H) extension $\mathbf{y}$ over $\mathbf{x}$. Here the conjunction is over $\mathbf{y}$ with the y_j distinct from each other and from the x_i. Further it may be that two $\mathbf{y}$ which are permutations of each other give the tautologically same event (E.g.: "x_1, y_1, y_2 is a triangle" and "x_1, y_2, y_1 is a triangle") in which case we agree to take just one $\mathbf{y}$ from each equivalence class. There are $\sim cn^v p^e$ different $\mathbf{y}$ – the constant c depending on the group of symmetries. Each $\Pr[B_\mathbf{y}] = p^e$ which is certainly $o(1)$. Thus

$$M \sim (1 - p^e)^{cn^v} \sim e^{-cn^v p^e} \sim e^{-cn^{e\epsilon}}$$

which rapidly goes to zero.

What about Δ? Let us make, for the moment, an additional assumption that ϵ is small, as described below. Write $\Delta = \sum \Pr[B_\mathbf{y} \wedge B_\mathbf{z}]$, the sum over $\mathbf{y} = (y_1, \ldots, y_v)$, $\mathbf{z} = (z_1, \ldots, z_v)$ so that the requisite extensions would have at least one common edge. The sum can be split into a finite number of types depending on the intersection pattern of $\mathbf{y}, \mathbf{z}$. First suppose $\mathbf{y}, \mathbf{z}$ have the same v vertices. Because we have excluded the case where $B_\mathbf{y}, B_\mathbf{z}$ are tautologically identical the event $B_\mathbf{y} \wedge B_\mathbf{z}$ must require at least $e + 1$ edges to be in G and hence has probability at most p^{e+1}. There are $\Theta(n^v)$ such pairs $\mathbf{y}, \mathbf{z}$ (we may think of v degrees of freedom here) and so their contribution to Δ is bounded from above by $\Theta(n^v p^{e+1}) = O(n^{(e+1)\epsilon - a})$. For ϵ small (dependent on e and v) this exponent is negative and so that contribution is $o(1)$.

Now suppose that $\mathbf{y}, \mathbf{z}$ have t coordinates in common, though the coordinates need not be in the same positions. We must have $t > 0$ as otherwise $\Pr[B_\mathbf{y} \wedge B_\mathbf{z}]$ would not be counted in Δ. Suppose their overlap has t vertices and e_t edges. (In counting overlap edges we count those $\{y_1, y_2\}$ where both of y_1, y_2 are coordinates of both $\mathbf{y}$ and $\mathbf{z}$ and those $\{x_i, y\}$ where x_i is a coordinate of $\mathbf{x}$ and y is a coordinate of both $\mathbf{y}$ and $\mathbf{z}$.) Then $B_\mathbf{y} \wedge B_\mathbf{z}$ is an extension with $2v - t$ vertices and $2e - e_t$ edges. There are $\Theta(n^{2v-t})$ such pairs $\mathbf{y}, \mathbf{z}$, each vertex of overlap costing us one degree of freedom. The contribution to Δ is then bounded from above by $O(n^{2v-t-(2e-e_t)(a-\epsilon)})$. We rewrite the exponent as $(2e - e_t)\epsilon - (t - e_t a)$. The assumption that (R, H) is strictly balanced tells us that $\frac{e_t}{t} < \frac{e}{v} = \frac{1}{a}$ so that $t - e_t a$ is positive. Thus for ϵ small (dependent on e, v, t) the exponent is negative and the contribution to Δ is $o(1)$.

We have a finite number of conditions on ϵ so we can pick ϵ small so that it satisfies them all. Then each of the finite number of types of overlap contribute $o(1)$ to Δ and hence $\Delta = o(1)$. This condition assures that Janson's

Inequality gives an asymptotic formula $\Pr[B] \sim M$. Finally, M, as calculated above, is subpolynomial, certainly $o(n^{-r})$ with r the number of roots. Thus $\Pr[\bigvee B]$, the disjunction taken over all choices of $\mathbf{x}$, is still $o(1)$. Then $\overline{\vee B}$ holds almost surely, and that is precisely the event $\mathrm{Ext}(R, H)$.

This completes the proof if ϵ is small. But $\mathrm{Ext}(R, H)$ is a monotone property and hence the larger p is the larger the probability that it is true. Thus if it is true for $p = n^{-a+\epsilon}$ with ϵ small and positive it will be true for $p = n^{-a+\epsilon}$ for any positive ϵ.

5.2 Counting Extensions

Let $p = n^{-\alpha}$ with $\alpha < \frac{2}{3}$. For each vertex x in $G(n, p)$ let N_x denote the number of triangles containing x. We have seen that almost surely all N_x are positive. A simple expected value argument gives $E[N_x] = \binom{n-1}{2}p^3 \sim \frac{1}{2}n^{2-3\alpha}$. Our object here is to show that almost surely every N_x is close to its expectation.

As before we have a more general framework. Let $p = n^{-\alpha}$ with $\alpha \in (0, 1)$. Let (R, H) be a *safe* extension with respect to α as given by Definition 4.4. Let (R, H) have type (v, e) with r roots. For any $\mathbf{x} = (x_1, \ldots, x_r)$ of distinct vertices in $G(n, p)$ let $N_{\mathbf{x}}$ denote the number of (R, H) extensions (as given by Definition 1.5) $\mathbf{y}$. Set $\mu = E[N_{\mathbf{x}}]$ so that $\mu \sim n^{v-e\alpha}$.

The following result is not trivial. We note, however, that it is in some sense the *only* difficult [depending on your definition!] result from random graphs that the nonprobabilist reading these pages need know. Indeed, such a reader may wish simply to read the result and move on to the next section. For the probabilist we note a recent paper of Kim and Vu [11]. That paper develops a powerful methodology which yields this result as a corollary.

Theorem 5.2.1 (Counting Extensions Theorem). *With (R, H) safe and α, p, v, e, r, μ and $N_{\mathbf{x}}$ as above almost surely*

$$N_{\mathbf{x}} \sim \mu$$

for all $\mathbf{x} = (x_1, \ldots, x_r)$ *of distinct vertices in* $G(n, p)$.

We begin with a reduction. Suppose the Counting Extensions Theorem holds for minimally safe (R, H) as given by Definition 4.5. Let (R, H) be safe. We decompose it, finding $R = R_0 \subset R_1 \subset \ldots \subset R_t = V(H)$ such that each (R_{i-1}, R_i) is safe. Any choice $\mathbf{x}$ of $R = R_0$ then extends to R_1, thence $R_2, \ldots,$ until reaching a $\mathbf{y}$ of $R_t = V(H)$ in asymptotically the same number of ways. Hence the result would hold for (R, H).

We illustrate the above with a simple example. Suppose $\frac{1}{2} < \alpha < \frac{2}{3}$. For $\mathbf{x} = (x_1, x_2)$ let $N_{\mathbf{x}}$ be the number of (y_1, y_2, z) such that $x_1 y_1 y_2 x_2$ is a path and z, y_1 are adjacent. Given the Counting Extensions Theorem for minimally safe extensions almost surely for every x_1, x_2 there would be $\sim n^2 p^3$ choices of y_1, y_2 with $x_1 y_1 y_2 x_2$ a path. But then for each such y_1 there would be $\sim np$ choices of z with z, y_1 a path. Thus $N_{\mathbf{x}} \sim n^3 p^4$, as desired.

We therefore assume (R, H) is minimally safe.

Claim 5.2.2 *Let (R, H) be minimally safe of type (v, e) with $v > 1$. Then $v - e\alpha < 1$.*

Proof: Pick any $x \in V(H) - R$. Then Property 4.1.16 gives that $(R \cup \{x\}, H)$ is dense. It has $v' = v - 1$ nonroots and at most e edges so $v - 1 - e\alpha < 0$.

Claim 5.2.3 *Let (R, H) be minimally safe and ϵ be positive and fixed. Let α, p, v, e, r, μ be as above. Let $\mathbf{x}$ be a fixed r-tuple from $G(n, p)$. Then*

$$\Pr\left[N_{\mathbf{x}} < (1 - \epsilon)\mu\right] = o(n^{-r}) \text{ and } \Pr\left[N_{\mathbf{x}} > (1 + \epsilon)\mu\right] = o(n^{-r})$$

Claim 5.2.3 implies Theorem 5.2.1 as there are only $O(n^r)$ possible $\mathbf{x}$. The actual probabilities are exponentially small which will give us, in a certain sense, plenty of room. We fix $\mathbf{x}$ and now aim for Claim 5.2.3.

Claim 5.2.4 $\Pr[N_{\mathbf{x}} = 0] = e^{-\mu(1+o(1))}$.

Proof: We apply Janson's Inequality as done earlier. For each $\mathbf{y}$ we have the event $B_{\mathbf{y}}$ that $\mathbf{y}$ is an (R, H) extension of $\mathbf{x}$. Then μ is as desired. It suffices to show $\Delta = o(\mu)$. We can split Δ into the bounded number of ways two (R, H) extensions $\mathbf{y}, \mathbf{z}$ can overlap so it suffices to show that the contribution from any particular overlap is $o(\mu)$. Consider an overlap with v_1 vertices and e_1 edges. Deleting the overlap the extension has $v - v_1$ vertices and $e - e_1$ edges. From Property 4.1.16 this extension is dense, $v - v_1 - \alpha(e - e_1)$ is negative. The contribution to Δ from these pairs is then n to the power $v + (v - v_1) - \big(e + (e - e_1)\big)\alpha$. The power is less than $v - e\alpha$ and so the contribution is $o(\mu)$. This completes Claim 5.2.4.

Definition 5.2 *A family F of (R, H) extensions $\mathbf{y}$ of $\mathbf{x}$ is called* disjoint *if no two $\mathbf{y}_1, \mathbf{y}_2 \in F$ have a common coordinate, even if the coordinates lie in different positions. It is called* maximal disjoint *if there is no F' with $F \subset F'$ which is also a disjoint family of (R, H) extensions.*

Claim 5.2.5 *Let (R, H) be minimally safe and ϵ be positive and fixed. Let α, p, v, e, r, μ be as above. Let $\mathbf{x}$ be a fixed r-tuple from $G(n, p)$. Then the probability that there is any maximal disjoint family F of (R, H) extensions of any size i where $i < (1 - \epsilon)\mu$ or $i > (1 + \epsilon)\mu$ is $o(n^{-r})$.*

Proof: As there are at most n possible values for i it suffices to show that this probability is $o(n^{-r-1})$. Fix i with $i > (1 + \epsilon)\mu$ or $i < (1 - \epsilon)\mu$. We shall actually show that the expected number of maximally disjoint families F of size i is exponentially small in n.

There are cn^v potential (R, H) extensions $\mathbf{y}$, each is an (R, H) extension with probability p^e and $\mu = cn^v p^e$. The expected number of $F = \{\mathbf{y}_1, \ldots, \mathbf{y}_i\}$

of disjoint (R,H) extensions is then at most $\binom{cn^v}{i}p^i \le \mu^i/i!$. For $i > 3\mu$ (actually, 3 can be any constant bigger than e) we use Stirling's Formula to bound this by $(e\mu/i)^{i(1+o(1))} < (e/3)^{\mu}$ which is exponentially small. For $i \le 3\mu$ we now invoke the maximality. Given such an F for it to be maximal there can be no (R,H) extension on the $n' = n - r - iv$ other vertices. From Claim 5.2.2 $\mu = O(n^{v-e\alpha}) \ll n$ so $n' \sim n$. We apply Claim 5.2.4 to give that the probability of there being no (R,H) extensions among those n' vertices is $e^{-\mu'(1+o(1))}$ where μ' is the expected number of such extensions. But since $n' \sim n$, $\mu' \sim \mu$ so this may be written $e^{-\mu(1+o(1))}$. Thus the expected number of F of size i is bounded by $e^{-\mu(1+o(1))}\mu^i/i!$. If we parametrize $i = y\mu$ and apply Stirling's Formula this becomes $(e^{y-1}y^{-y})^{\mu(1+o(1))}$. The function $e^{y-1}y^{-y}$ hits its maximum value 1 at $y = 1$. For $y < 1-\epsilon$ and $1+\epsilon < y < 3$ we therefore have an exponentially small upper bound. This concludes Claim 5.2.5.

Tautologically, there must be some maximal F. Thus with probability $1 - o(n^{-r})$ there is a maximal F of size between $(1-\epsilon)\mu$ and $(1+\epsilon)\mu$. As all $\mathbf{y} \in F$ are (R,H) extensions we have that $N_{\mathbf{x}} \ge \mu(1-\epsilon)$. For the upper bound on $N_{\mathbf{x}}$ we need show that $N_{\mathbf{x}}$ cannot be much greater than $|F|$. Let F^* be the set of all (R,H) extensions $\mathbf{y}$ of $\mathbf{x}$. Consider F^* as a graph, with $\mathbf{y}_1, \mathbf{y}_2$ adjacent if they share a common vertex.

We first bound the degree of F^*. (This part of the argument is very similar to the proof of the Finite Closure Theorem 4.3.2.) Let $\epsilon > 0$ be such that $v - \alpha e \le -\epsilon$ for all positive integers v', e' with $v \le v$ and $v - \alpha e$ negative. Pick K with $r+1-K\epsilon$ negative. Set $K_1 = (vK-1)_v$. Let w be a vertex of $G(n,p)$. If w were in more than K_1 extensions $\mathbf{y}$ then the extensions would use at least vK vertices and so there would be extensions $\mathbf{y}_1, \ldots, \mathbf{y}_K$, each using at least one new vertex. Consider starting with $\mathbf{x}$ and w. Each extension adds some v_i vertices and e_i edges with (by Claim 5.2.2) $v_i - \alpha e_i < 0$. Thus all $v_i - \alpha e_i \le -\epsilon$. The subgraph consisting of $\mathbf{x}$, w and the extensions would then have $V = r + 1 + \sum v_i$ vertices and at least $E = \sum e_i$ edges. But $V - \alpha E \le r + 1 - K\epsilon$ is negative. Then almost surely for no choice of $\mathbf{x}$ do there exist K_1 extensions $\mathbf{y}$.

Now we bound the number of independent edges. There are only a finite number of ways no extensions $\mathbf{y}_1, \mathbf{y}_2$ can intersect so we may fix one way. Their union would then have V vertices and E edges where $V - \alpha E < v - \alpha e$, as described before. Set μ_1 equal the expected number of such extensions, so that $\mu_1 \ll \mu$, having a lower power of n. The probability of having $3\mu_1$ disjoint such extensions is then exponentially small.

Now let F be the family of disjoint $\mathbf{y}$ of maximal size and suppose there are L other extensions $\mathbf{y}$. Each other $\mathbf{y}$ is adjacent to some $\mathbf{y}' \in F$. Let K_1 be the bound on the degree of any $\mathbf{y}'$. Suppose $L > K_1(3\mu_1)$. We select $\mathbf{y} \notin F$ and then an adjacent $\mathbf{y}' \in F$ and then delete all other $\mathbf{y}$ adjacent to $\mathbf{y}'$. Continuing in this way we would find $3\mu_1$ independent edges, which does not occur. Hence $L \le 3K_1\mu_1 \ll \mu$. That is, the total number of extensions $\mathbf{y}$ is $\sim \mu$. This completes Claim 5.2.3 and hence Theorem 5.2.1.

One further extension will be useful in the next section.

Theorem 5.2.6. *With (R, H) safe and α, p, v, e, r, μ as in Theorem 5.2.1 and K fixed almost surely for all $\mathbf{x} = (x_1, \ldots, x_r)$ of distinct vertices in $G(n,p)$ and every K other vertices $w_1, \ldots, w_K$ the number of (R, H) extensions $\mathbf{y}$ that do not include any of the z_j is asymptotic to μ.*

Proof: There are $O(n^{r+K})$ choices for $\mathbf{x}$ and the z_j. Hence it suffices to show Claim 5.2.3 with failure probability $o(n^{-r-K})$ on the graph $G(n,p)$ with the z_j removed. The proof of Claim 5.2.3 gave failure probability exponentially small. The graph $G(n,p)$ with the z_j removed is the random graph $G(n-K, p)$ so that all calculations are as before with the K having a negligible effect.

5.3 Generic Extension

Let α positive be fixed, let (R, H) be a rooted graph, and let t be a positive integer.

Definition 5.3 *We say* $\mathbf{y} = (y_1, \ldots, y_v)$ *is a t-*generic (R, H) extension of $\mathbf{x}$ *if*
- $\mathbf{y}$ *is an (R, H) extension of* $\mathbf{x}$
- *There are no additional edges between the y_j or from the y_j to the x_i beyond the requirement of a (R, H) extension*
- *If any $\mathbf{z} = (z_1, \ldots, z_s)$ with $s \le t$ forms a rigid extension over H then there are no edges between the z_k and the y_j.*

The notion here is that $\mathbf{y}$ forms an (R, H) extension over $\mathbf{x}$ but has no other relationship with $\mathbf{x}$.

Theorem 5.3.1 (Generic Extension Theorem). *If (R, H) is safe, relative to α, and $G \sim G(n,p)$ with $p = n^{-\alpha}$ then almost surely every $\mathbf{x}$ has a t-generic (R, H) extension $\mathbf{y}$.*

Proof: Let (R, H) have type (v, e). For any $\mathbf{x}$ let $w_1, \ldots, w_L$ be the vertices of $\mathrm{cl}_{t+u}(\mathbf{x}) - \mathbf{x}$. By the Finite Closure Theorem 4.3.2 we may bound $L \le K$ where K depends only on $r, t+u$ and α but not on the choice of $\mathbf{x}$. From Theorem 5.2.6 there are $\Theta(n^v p^e)$ (R, H) extensions $\mathbf{y}$ using none of the ws.

We bound from above the number of such $\mathbf{y}$ that fail to be t-generic. It may be that there are additional edges $x_i y_j$ or $y_j y_k$. Then $\mathbf{y}$ would form an (R, H^+) extension of $\mathbf{x}$ of type (v, e^+) with $e^+ > e$. This could have no rigid subextensions as no $y_i = w_j$. Hence, by Property 4.1.7, (R, H^+) would be safe. By Theorem 5.2.6 the would be $O(n^v p^{e^+})$ such $\mathbf{y}$. But $O(n^v p^{e^+}) = o(n^v p^e)$.

Further, there might be a rigid extension over H with at most t nonroots. If so some $\mathbf{z}$ would form a rigid extension (H, H_1) over H which was minimal in the sense that no subextension was rigid. Let (H, H_1) have type (v_1, e_1) so that $v_1 - e_1\alpha < 0$. Suppose (R, H_1) is safe. Each $\mathbf{x}$, by Theorem 5.2.6, has $O(n^{v+v_1}p^{e+e_1})$ extensions to H_1 and hence at most that many $\mathbf{y}$. But $O(n^{v+v_1}p^{e+e_1}) = o(n^v p^e)$.

Finally, suppose (R, H_1) is not safe. By Property 4.1.7 there would be a rigid subextension (R, H_2). Then $V(H_2) \subseteq \mathrm{cl}_t(R)$ so H_2 has no ys. From Property 4.1.12 $\big(V(H), V(H_2) \cup (V(H) - R)\big)$ is rigid. By minimality $V(H_2) \cup \big(V(H) - R\big)$ is all of $V(H_1)$. That is, $V(H_2)$ consists precisely of the xs and the zs, $\mathbf{z}$ is a rigid extension over $\mathbf{x}$.

Consider $\mathbf{y}$ as an extension of $\mathbf{x}, \mathbf{z}$. The extension has type (v, e^+) with $e^+ > e$ since, critically, there must be at least one edge $z_k y_j$. As argued earlier there can be no rigid subextensions so the extension would be safe and there would be $O(n^v p^{e^+}) = o(n^v p^e)$ such $\mathbf{y}$. Given $\mathbf{x}$ there are $O(1)$ $\mathbf{z}$ by the Finite Closure Theorem 4.3.2 and then $o(n^v p^e)$ $\mathbf{y}$, giving a total of $o(n^v p^e)$ $\mathbf{y}$.

$\mathbf{y}$ can fail to be t-generic in only a bounded number of ways and we have seen that for each way there are $o(n^v p^e)$ such $\mathbf{y}$. Hence there is at least one $\mathbf{y}$ – indeed, $\Omega(n^v p^e)$ such $\mathbf{y}$ – that are t-generic.

6. The Main Theorem

We fix an irrational $\alpha \in (0,1)$ throughout this chapter. All probabilities are with respect to the random graph $G(n,p)$ with $p = n^{-\alpha}$. We recall the statement of our goal, the Main Theorem 1.4.1: For any first order A

$$\lim_{n\to\infty} \Pr\big[G(n,n^{-\alpha}) \models A\big] = 0 \text{ or } 1$$

6.1 The Look-Ahead Strategy

Our approach is through the Ehrenfeucht Game as described in Section 2.1. We fix the number k of moves. We shall give a strategy for Duplicator so that, as $n,m \to \infty$, she almost surely wins $\textsc{Ehr}(G_1, G_2, k)$ where $G_1 \sim G(n,n^{-\alpha})$ and $G_2 \sim G(m,m^{-\alpha})$ are independently chosen.

Let $0 = t_0, t_1, \ldots t_{k-1}$ be nonnegative integers. The $(t_0, \ldots, t_{k-1})-$ *look-ahead strategy* for Duplicator is easy to describe. Duplicator makes any moves in response to Spoiler so that when there are i rounds remaining in the game the t_i-types of the vertices chosen are the same in both graphs. That is, if $x_1, \ldots, x_{k-i} \in G_1$, $y_1, \ldots, y_{k-i} \in G_2$ have been chosen then there is a graph isomorphism from $\mathrm{cl}_{t_i}(x_1, \ldots, x_{k-i})$ to $\mathrm{cl}_{t_i}(y_1, \ldots, y_{k-i})$ sending each x_j to its corresponding y_j.

Of course, it may well be that Duplicator is unable to keep to this strategy. In that case she loses. But if she is able to keep to this strategy then at the end of the game the 0-closures are the same and she has won. We shall give explicit (though surprisingly complicated) $t_0, \ldots, t_{k-1}$ so that Duplicator shall almost surely be able to keep to this strategy. Formally, we find t_i by induction on i. Note that as i represents the number of remaining moves we are really working backwards from the end of the game. We need show that almost surely for *every* $x_1, \ldots, x_{k-i} \in G_1$ and $y_1, \ldots, y_{k-i} \in G_2$ that have the same t_i-type and every $x_{k-i+1} \in G_1$ (Spoiler move) there exists $y_{k-i+1} \in G_2$ (Duplicator move) so that the resulting $k-i+1$-tuples have the same t_{i-1}-type. [Of course, Spoiler could also move in G_2 but this case is the same by symmetry.] For convenience of exposition we consider the first $(i = k)$ and final $(i = 1)$ moves separately. In a formal sense this is unnecessary.

6.1.1 The Final Move

We set $t_1 = 1$. Assume $x_1, \ldots, x_{k-1} \in G_1$, $y_1, \ldots, y_{k-1} \in G_2$ have been chosen with the same 1-closure. Now Spoiler moves and by symmetry we can assume he picks $x_k \in G_1$. (Note that we cannot assume x_k is a random choice, quite the opposite!) Let w be the number of previously selected x's adjacent to the newly selected x_k.

Case 1 (Inside): $1 - w\alpha < 0$. Then $x_k \in \mathrm{cl}_1(x_1, \ldots, x_{k-1})$ since the rooted graph with $v = 1$ nonroot and w edges is rigid. Let $\Psi\colon \mathrm{cl}_1(x_1, \ldots, x_{k-1}) \to \mathrm{cl}_1(y_1, \ldots, y_{k-1})$ be the isomorphism guaranteed by the 1-types being the same. Spoiler selects $y_k = \Psi(x_k)$. (Wily Spoiler's attempt to trick Duplicator, as in Section 2.5, is thwarted by her having looked ahead and assured that not only the induced graphs but the 1-closures were identical.)

Case 2 (Outside): $1 - w\alpha > 0$. The rooted graph with $k - 1$ roots and one nonroot adjacent to w of the roots is now safe. By our Generic Extension Theorem 5.3.1 almost surely for every $k - 1$ vertices in G_2 there is a vertex adjacent to any prescribed w of them and no others. Duplicator picks that $y_k \in G_2$ adjacent to just those w of the $y_j \in G_2$ such that the x_k selected by Spoiler was adjacent in G_1 to x_j.

6.1.2 The Core Argument (Middle Moves)

Let us fix i with $1 \le i \le k - 1$ and let t_i be given. We select t_{i+1} so that

1. $t_{i+1} \ge t_i$
2. Almost surely the t_i-closure of any $k - i$ vertices has at most $t_{i+1} - 1$ nonroots.

The existence of such t_{i+1} is a consequence of the Finite Closure Theorem 4.3.2. For notational convenience we set $t = t_i$, $u = t_{i+1}$. Assume $x_1, \ldots, x_{k-i-1} \in G_1$, $y_1, \ldots, y_{k-i-1} \in G_2$ have been chosen with the same u-closure. Set $\mathbf{x} = (x_1, \ldots, x_{k-i-1})$, $\mathbf{y} = (y_1, \ldots, y_{k-i-1})$ for further notational convenience. Let $\Psi\colon \mathrm{cl}_u(\mathbf{x}) \to \mathrm{cl}_u(\mathbf{y})$ be the graph isomorphism showing that their u-types are the same. Now Spoiler selects $x \in G_1$.

Case 1 (Inside) $x \in \mathrm{cl}_u(\mathbf{x})$. Spoiler selects $y = \Psi(x)$. As $t \le u$ the u-closure of $\mathbf{x}, x$ is contained in $\mathrm{cl}_u(\mathbf{x})$ (and also for $\mathbf{y}$) so that restricting Ψ gives an isomorphism from $\mathrm{cl}_t(\mathbf{x}, x)$ to $\mathrm{cl}_t(\mathbf{y}, y)$.

Case 2 (Outside) $x \notin \mathrm{cl}_u(\mathbf{x})$. Set $H = \mathrm{cl}_t(\mathbf{x}, x)$ and $R = \mathrm{cl}_t(\mathbf{x}, x) \cap \mathrm{cl}_u(\mathbf{x})$ and consider the rooted graph (R, H). As $x \in H$ but by assumption $x \notin R$ this is a legitimate extension.

We claim (R, H) is safe. Otherwise by Property 4.1.7 it would have a rigid subextension (R, H'). The number of nonroots of (R, H') would be at most the number of nonroots of $(\mathbf{x}, \mathrm{cl}_t(\mathbf{x}, x))$ which is one plus the number of nonroots of $\big((\mathbf{x}, x), \mathrm{cl}_t(\mathbf{x}, x)\big)$. We've designed $u = t_{i+1}$ so that this is at most $1 + (u - 1) = u$. Then

$$H' \subseteq \mathrm{cl}_u(R) \subseteq \mathrm{cl}_u(\mathrm{cl}_u(\mathbf{x})) = \mathrm{cl}_u(\mathbf{x}) \subseteq R$$

a contradiction.

Set $R' = \Psi(R)$. We apply Generic Extension Theorem 5.3.1 to give a u-generic (R, H) extension over R', call it (R', H'). The isomorphism Ψ, limited to $R \to R'$, extends to a graph isomorphism $\Psi^+ \colon H \to H'$. Spoiler selects $y = \Psi^+(x)$.

Does this work? As H, H' are isomorphic $\mathrm{cl}_t(\mathbf{y}, y)$ certainly contains H'. There are no additional edges in H' since there were none in R' and the extension was generic. Can there be more points in $\mathrm{cl}_t(\mathbf{y}, y)$? Set $\mathrm{NEW}' = \mathrm{cl}_t(\mathbf{y}, y) - H'$. If $\mathrm{NEW}' \neq \emptyset$ then $\mathrm{cl}_t(\mathbf{y}, y)$ would be a rigid extension over H' and, by the general bound, NEW' would have at most u vertices. Thus the $v \in \mathrm{NEW}'$ would be adjacent only to vertices in $H' \cap \mathrm{cl}_u(\mathbf{x})$ (not to $H' - \mathrm{cl}_u(\mathbf{x})$) and would be a rigid extension over $H' \cap \mathrm{cl}_u(\mathbf{x})$. Hence $\mathrm{NEW}' \subset \mathrm{cl}_u(\mathrm{cl}_u(\mathbf{x}))$ which is simply $\mathrm{cl}_u(\mathbf{x})$. But then back in G_1, setting $\mathrm{NEW} = \Psi^{-1}(\mathrm{NEW}')$ we have the same picture with $H \cup \mathrm{NEW}$ isomorphic to $H' \cup \mathrm{NEW}'$. In G_2 all points of $H' \cup \mathrm{NEW}'$ can be reached from $\mathbf{y}, y$ by rigid extensions of at most t nonroots and those extensions never go outside of $H' \cup \mathrm{NEW}'$. But then the same would be true in G_1 with the isomorphic $H \cup \mathrm{NEW}$ and that would give $\mathrm{cl}_t(\mathbf{x}, x)$ extra vertices that it doesn't have.

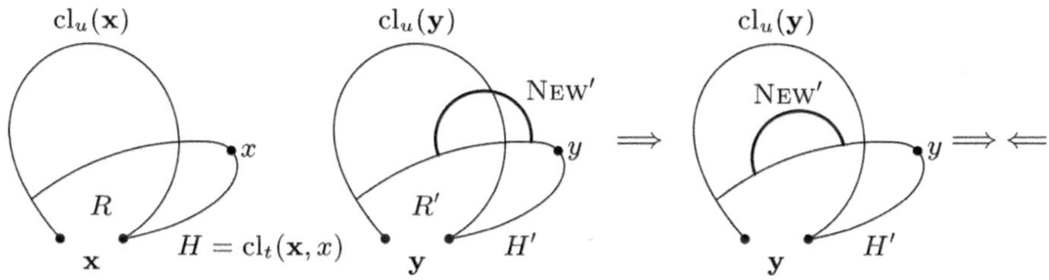

Fig. 6.1. Spoiler plays Outside x. Duplicator finds y with u-generic (R, H)-extension. If $\mathrm{cl}_t(\mathbf{y}, y)$ had other vertices NEW' the genericity would force them inside $\mathrm{cl}_u(\mathbf{y})$ but then they would have been in $\mathrm{cl}_t(\mathbf{x}, x)$ as well

6.1.3 The First Move

Set $t = t_k$. On the first move Spoiler selects some $x \in G_1$. Duplicator calculates $\mathrm{cl}_t(x)$ and must find an $y \in G_2$ with the same t-type. To show that she almost surely succeeds one needs that every t-type of a single vertex either almost surely or almost never appears in $G(n, n^{-\alpha})$. There are only a finite number (by the Finite Closure Theorem 4.3.2) of possible t-types to consider so it suffices to show this for any particular one. We write the t-type as the graph $H = \mathrm{cl}_t(x)$, with vertex x specified. We look only at logically possible H, so that $(\{x\}, H)$ is a rigid extension.

Suppose H contains a subgraph H_1 with v_1 vertices, e_1 edges where $v_1 - e_1\alpha < 0$. The expected number of copies of H_1 is $O(n^{v_1}p^{e_1}) = O(n^{v_1 - e_1\alpha})$

which is $o(1)$ so almost surely there are no copies of H_1 and hence no copies of H and hence no x with $H = \mathrm{cl}_t(x)$ exists.

Otherwise $(\emptyset, H)$ is a safe extension. Then not only does there exist a copy of H but by the Generic Extension Theorem 5.3.1 there exists an induced copy of H which is t-generic over the empty set – which means that $\mathrm{cl}_t(H) = H$. We know $\mathrm{cl}_t(x)$ contains H, but it is also contained in $\mathrm{cl}_t(H) = H$ and therefore it is precisely H.

6.2 The Original Argument

We begin by restating the crucial idea of Section 6.1.2 which in some sense is the centerpiece of the entire argument.

Theorem 6.2.1. *Let $u \geq t$ be such that almost surely the t-closure of any $k+1$ vertices has at most $u-1$ nonroots. Let H be any possible value of $\mathrm{cl}_u(\mathbf{x})$, where we set $\mathbf{x} = (x_1, \ldots, x_k)$. Let H_1 be any possible value of $\mathrm{cl}_t(\mathbf{x}, x)$. Then almost surely either*
- *For every $\mathbf{x}$ with $\mathrm{cl}_u(\mathbf{x}) \cong H$ there exists x with $\mathrm{cl}_t(\mathbf{x}, x) \cong H_1$*

or
- *For every $\mathbf{x}$ with $\mathrm{cl}_u(\mathbf{x}) \cong H$ there does* not *exist x with $\mathrm{cl}_t(\mathbf{x}, x) \cong H_1$.*

We want H, H_1 to have common vertices $x_1, \ldots, x_k$. They may or may not have other common vertices. We call H^* a *picture* if it is derived from H, H_1 by identifying the roots (in the prescribed order) and identifying some (possibly none) other pairs of vertices and otherwise keeping the vertices distinct. As H^* has bounded size there are only a finite number of possible pictures H^*. We actually show that for every such H^* almost surely either
- For every $\mathbf{x}$ with $\mathrm{cl}_u(\mathbf{x}) \cong H$ there exists x with $\mathrm{cl}_t(\mathbf{x}, x) \cong H_1$ and $H \cup H_1 \cong H^*$

or
- For every $\mathbf{x}$ with $\mathrm{cl}_u(\mathbf{x}) \cong H$ there does not exist x with $\mathrm{cl}_t(\mathbf{x}, x) \cong H_1$ and $H \cup H_1 \cong H^*$.

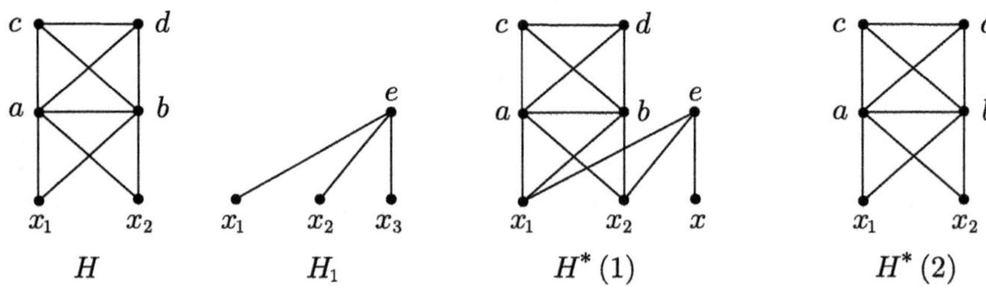

Fig. 6.2. $\alpha = \pi/7$. $H \cong \mathrm{cl}_2(x_1, x_2)$. $H_1 \cong \mathrm{cl}_1(x_1, x_2, x)$. $H^*(1)$, $H^*(2)$ are two (of many) possible pictures given by identification. In $H^*(2)$ x is Inside. But in $H^*(2)$ $a, c \in \mathrm{cl}_1(x_1, x_2, x)$, hence no x can exist with this picture. In $H^*(1)$ x is Outside. $\{x, e\}$ is safe over $H_0 = \{x_1, x_2\}$. By Generic Extension such x, e almost surely exist

In the picture H^* set $H_0 = H \cap H_1$. There are two cases.

Case 1 (Inside) $x \in H_0$. Then $H = \mathrm{cl}_u(\mathbf{x})$ determines $\mathrm{cl}_t(\mathbf{x}, x)$ so we must have $H_1 \subseteq H$ and can check if H_1 is indeed $\mathrm{cl}_t(\mathbf{x}, x)$.

Case 2 (Outside) $x \notin H_0$. Then we must have (H_0, H_1) safe. Otherwise by Property 4.1.7 it would have a rigid subextension (H_0, H_2) whose number of nonroots would be at most $u - 1 + 1 = u$ and then H_2 would have to be contained in H. It further must be that in H^* the t-closure of $\mathbf{x}, x$ is H_1, nothing more or less. But suppose these are satisfied. By the Generic Extension Theorem 5.3.1 for all $\mathbf{x}$ with $\mathrm{cl}_u(\mathbf{x}) \cong H$ there will exist a u-generic extension giving H^*. If the t-closure of $\mathbf{x}, x$ was H_2, strictly more than H_1, then H_2 would be a rigid extension over H_1 but by genericity it would be a rigid extension over H_0 but then it would be in H which we have already checked.

The original proof of the Main Theorem 1.4.1 did not use the Ehrenfeucht game. Rather, it was an induction on the length of the statement. To make the induction go, however, we need to prove a statement for all predicates $P(x_1, \ldots, x_k)$ with any number k of free variables. Sentences have $k = 0$ free variables. As before, α is a fixed irrational number between zero and one.

Theorem 6.2.2. *For every predicate $P(x_1, \ldots, x_k)$ there exists a nonnegative integer t so that the following holds almost surely in $G(n,p)$ with $p = n^{-\alpha}$: for each t-type H either all $x_1, \ldots, x_k$ with that t-type satisfy P or no $x_1, \ldots, x_k$ with that t-type satisfy P.*

The proof is by induction on the length of the predicate P. For the atomic predicates $x_i \sim x_j$ and $x_i = x_j$ we can take $t = 0$ as the 0-closure includes this information. If the statement holds for P then it certainly holds for $\neg P$ with the same value of t. If the statement holds for P, Q with values t_1, t_2 then it holds for $P \wedge Q$ (or any Boolean function of P, Q) with the value $t = \max(t_1, t_2)$, as a t-type includes the information about the s-type for all smaller s. This leaves us with the one important case, a predicate of the form $Q = \exists_x P(x_1, \ldots, x_k, x)$. By induction there is a t so the $P(x_1, \ldots, x_k, x)$ holds if and only if the t-type of $x_1, \ldots, x_k, x$ is one of some finite list $H_1, \ldots, H_s$. Let u satisfy the conditions of Theorem 6.2.1. There is a finite list of potention u-types for $x_1, \ldots, x_k$. For each such u-type, either almost surely there exists x with $x_1, \ldots, x_k, x$ having t-type one of $H_1, \ldots, H_s$ or almost surely no such x exists. Call those u-types positive and negative respectively. Then almost surely Q holds if and only if the u-type of $x_1, \ldots, x_k$ is positive.

7. Countable Models

Again we fix an irrational α between zero and one. Since the random graphs $G(n,p)$ with $p = n^{-\alpha}$ satisfy the Zero-One Law of our Main Theorem 1.4.1 the set T_α of sentences that hold almost always form a complete theory. Here we examine an axiomatization of T_α and its countable models.

7.1 An Axiomatization for T_α

7.1.1 The Schema

The theory T_α can be axiomatized with two schema – a negative schema preventing subgraphs that shouldn't exist and a positive schema giving extensions that should exist.

NonExistence Schema [For every H with v vertices, e edges, $v - e\alpha < 0$] There does not exist a copy of H.

Generic Extension Schema [For every safe (R, H) and every $t \geq 0$] For all $\mathbf{x} = (x_1, \ldots, x_r)$ there exist $\mathbf{y} = (y_1, \ldots, y_v)$ forming a t-generic (R, H) extension over $\mathbf{x}$.

The NonExistence sentences have particularly simple form, being negations of purely existential statements. The Generic Extension sentences have a somewhat more complicated structure. They begin $\forall_{\mathbf{x}} \exists_{\mathbf{y}}$ so that the adjacencies $x_i y_j$ and $y_j y_k$ are completely set. Then further there do not exist $z_1, \ldots, z_t$ forming any of the finite number of rigid extensions over $\mathbf{x} \cup \mathbf{y}$ that would destroy t-genericity.

7.1.2 Completeness Proof

Theorem 7.1.1. *T_α is generated by the NonExistence Schema and Generic Extension Schema given above.*

Let S_α denote the theory generated by the NonExistence Schema and Generic Extension Schema. The axioms of both schema all hold almost surely in T_α. Hence S_α will be subtheory of T_α.

Now we parrot the proof of the Main Theorem 1.4.1 given in Section 6.2 inside S_α. This takes some care as, for example, $y \in \mathrm{cl}_t(\mathbf{x})$ is not a priori a first order predicate. Let H be a graph with distinguished roots $a_1, \ldots, a_r$ and nonroots $b_1, \ldots, b_v$. Suppose further the t-closure of the roots is all of H. For such H the statement "$\mathrm{cl}_t(\mathbf{x})$ contains H" is the first order predicate that there exist $y_1, \ldots, y_v$ so that the adjacencies of $x_1, \ldots, x_r, y_1, \ldots, y_v$ precisely mimic those of $a_1, \ldots, a_r, b_1, \ldots, b_v$. The statement $\mathrm{cl}_t(\mathbf{x}) \cong H$ is the existence of those $y_1, \ldots, y_v$ with the further proviso that for no $1 \le i \le t$ do there exist $z_1, \ldots, z_i$ with $\mathbf{z}$ giving a rigid extension over $\mathbf{x}, \mathbf{y}$. We emphasize that for H, t fixed this is only a finite set of conditions so that $\mathrm{cl}_t(\mathbf{x}) \cong H$ is indeed a first order predicate.

We first give the Finite Closure Theorem 4.3.2.

Claim 7.1.2 *For every r, t there exists a finite set $H_1, \ldots, H_s$ so that, setting $\mathbf{x} = (x_1, \ldots, x_r)$,*

$$\forall_{\mathbf{x}} \bigvee_{i=1}^{s} \mathrm{cl}_t(\mathbf{x}) \cong H_i$$

is a theorem of S_α.

Proof: Tautologically $\mathrm{cl}_t(\mathbf{x})$ contains H implies either $\mathrm{cl}_t(\mathbf{x}) \cong H$ or one of a finite number of statements $\mathrm{cl}_t(\mathbf{x})$ contains H^+, where H^+ is given by a single rigid extension of at most t vertices over H. In S_α we can spread this tree of implications to get a disjunction of $\mathrm{cl}_t(\mathbf{x}) \cong H_i$ and a further disjunction of $\mathrm{cl}_t(\mathbf{x})$ contains H_i^+. By taking this out to K nonroots all of the possibilities $\mathrm{cl}_t(\mathbf{x})$ contains H_i^+ will be eliminated by the NonExistence Schema.

Now we give Theorem 6.2.1 in S_α.

Claim 7.1.3 *Let $u \ge t$ be such that in S_α the t-closure of any $k+1$ vertices has at most $u-1$ nonroots. Let H be any possible value of $\mathrm{cl}_u(\mathbf{x})$, where we write $\mathbf{x} = (x_1, \ldots, x_k)$. Let H_1 be any possible value of $\mathrm{cl}_t(\mathbf{x}, x)$. Then either*

$$\forall_{\mathbf{x}}\, \mathrm{cl}_u(\mathbf{x}) \cong H \rightarrow \exists_x\, \mathrm{cl}_t(\mathbf{x}, x) \cong H_1$$

or

$$\forall_{\mathbf{x}}\, \mathrm{cl}_u(\mathbf{x}) \cong H \rightarrow \neg\exists_x\, \mathrm{cl}_t(\mathbf{x}, x) \cong H_1$$

is a theorem of S_α.

Proof: As in Theorem 6.2.1 we call H^* a *picture* if it is derived from H, H_1 by identifying the roots (in the prescribed order) and identifying some (possibly no) other pairs of vertices and otherwise keeping the vertices distinct. As H^* has bounded size there are only a finite number of possible pictures H^*. Let the first order predicate $\mathrm{pict}(H, H_1) = H^*$ list the identifications and nonidentifications. We actually show that for every such H^* either

$$\forall_{\mathbf{x}}\, \mathrm{cl}_u(\mathbf{x}) \cong H \rightarrow \exists_x \big[\mathrm{cl}_t(\mathbf{x}, x) \cong H_1 \wedge \mathrm{pict}(H, H_1) = H^*\big]$$

or

$$\forall_{\mathbf{x}}\, \mathrm{cl}_u(\mathbf{x}) \cong H \rightarrow \neg\exists_x \big[\mathrm{cl}_t(\mathbf{x}, x) \cong H_1 \wedge \mathrm{pict}(H, H_1) = H^*\big]$$

is a theorem of S_α. But now the proof of the Main Theorem 1.4.1 given in Section 6.2 goes through completely – the Inside case being an examination of cases and the Outside case invoking the Generic Extension schema in the positive cases.

Claim 7.1.4 *For every predicate* $P(\mathbf{x})$ *there is an integer* t *and a finite list* $H_1, \ldots, H_s$ *so that*

$$\forall_x \left[P(\mathbf{x}) \leftrightarrow \bigvee_{i=1}^{s} \mathrm{cl}_t(\mathbf{x}) \cong H_i\right]$$

is a theorem of S_α

Proof: We use induction on the length of P. For the atomic $P(x_1, x_2)$: $x_1 \sim x_2$ or $x_1 = x_2$ we use $t = 0$. For $\neg P(\mathbf{x})$ we make the disjuntion over all possible $\mathrm{cl}_t(\mathbf{x})$ (using Claim 7.1.2) that weren't listed for P. For $t' \leq t$ we have for any t-closure H that $\mathrm{cl}_t(\mathbf{x}) \cong H \rightarrow \mathrm{cl}_{t'}(\mathbf{x}) \cong H_1$ with a determined H_1. Reversing (again using Claim 7.1.2) $\mathrm{cl}_{t'}(\mathbf{x}) \cong H_1$ is equivalent to a finite disjunction of statements $\mathrm{cl}_t(\mathbf{x}) \cong H$. Now for $P(\mathbf{x}) \vee Q(\mathbf{x})$. By induction each is equivalent to a finite disjunction of statements $\mathrm{cl}_t(\mathbf{x}) \cong H$ where t is the maximum of t_1, t_2 and the result holds for P, Q with, respectively, t_1 and t_2-closure. Then $P(\mathbf{x}) \vee Q(\mathbf{x})$ is logically equivalent to a disjunction over all values H that appear in the two lists. [$P(\mathbf{x}) \wedge Q(\mathbf{x})$ is a disjunction over the common values H. Formally, this is not necessary as $P \wedge Q = \neg(\neg P \vee \neg Q)$ is redundant.]

The core of the argument is for predicates $\exists_x P(\mathbf{x}, x)$. We let t be such, by induction, that $P(\mathbf{x}, x)$ is equivalent to a disjunction of sentences $\mathrm{cl}_t(\mathbf{x}, x) \cong H_1$. As we've already handled disjunctions, let us imagine $P(\mathbf{x}, x)$ is of the form $\mathrm{cl}_t(\mathbf{x}, x) \cong H$. Let u be as in Claim 7.1.3. From Claim 7.1.2 $\bigvee \mathrm{cl}_u(\mathbf{x}) \cong H_i$ is a theorem of S_α for a finite list $H_1, \ldots, H_{s'}$. Apply Claim 7.1.3 to each of these H_i. Then $\exists_x P(\mathbf{x}, x)$ is equivalent to the disjunction of the predicates $\mathrm{cl}_u(\mathbf{x}) \cong H_i$ over those i for which

$$\forall_{\mathbf{x}}\, \mathrm{cl}_u(\mathbf{x}) \cong H_i \rightarrow \exists_x\, \mathrm{cl}_t(\mathbf{x}, x) \cong H$$

is a theorem of S_α.

For sentences A Claim 7.1.4 gives that A is equivalent to a disjunction union of statements $\mathrm{cl}_t(\emptyset) \cong H$. But the NonExistence schema gives immediately that $\mathrm{cl}_t(\emptyset) \cong \emptyset$. If $H = \emptyset$ is in the list then A is a theorem in S_α and if it is not in the list then $\neg A$ is a theorem in S_α. As A was arbitrary S_α is complete and hence is T_α, completing the proof of Theorem 7.1.1.

7.1.3 The Truth Game

We know that any A holds either almost surely or almost never in $G(n, n^{-\alpha})$. Can we tell which one it is? For one thing, we have to have α in a nice form. Let us say α is good if there is a Turing Machine that decides if $a/b < \alpha$ for integer inputs a, b.

Theorem 7.1.5. *Suppose α is a positive irrational for which there is a decision procedure to tell if $a/b < \alpha$ for integer inputs a, b. Then there is a decision procedure to tell if A holds almost surely or almost never in $G(n, n^{-\alpha})$.*

Proof: The axiom schema for T_α depend only on whether various $v - e\alpha$ are positive or negative and hence form a recursive set. With input A we may search simultaneously for proofs in T_α of A and of $\neg A$. If we find a proof of A then A holds almost surely; if we find a proof of $\neg A$ then A holds almost never.

While interesting, Theorem 7.1.5 does not give us a good sense of the complexity of deciding whether $A \in T_\alpha$. Let us fix the quantifier depth k of A and let $t_0 = 0, t_1 = 1, \ldots, t_k$ be given by the look-ahead strategy. We write A in prenex normal form

$$A\colon Q_{x_1} \cdots Q_{x_k} P(x_1, \ldots, x_k)$$

where each Q is either existential ($\exists$) or universal ($\forall$) quantification and P is a Boolean combination of the atomic $x_i \sim x_j$.

Now we define a perfect information game we call the Truth Game. There are two players, Spoiler and Verifier. There is a board which has $k+1$ levels, numbers $0, \ldots, k$. On level 0 there is a single point, labelled $\emptyset$. On level k are all possible labelled graphs H on $x_1, \ldots, x_k$ – equivalently, all truth settings of the $\binom{k}{2}$ atomic $x_i \sim x_j$. We mark H Yes if P holds on H and No if it does not. On level i are all possible t_{k-i}-closures of $x_1, \ldots, x_i$. When H is on level i and H' is on level $i+1$ we say the pair (H, H') is an acceptable move if and only if

$$\forall_{x_1,\ldots,x_i} \operatorname{cl}_{t_{k-i}}(x_1, \ldots, x_i) \cong H \Rightarrow \exists_{x_{i+1}} \operatorname{cl}_{t_{k-i-1}}(x_1 \ldots, x_{i+1}) \cong H'$$

is a theorem of T_α.

At the start of the game a chip is placed on the point $\emptyset$ at level 0. There are k moves to the game. At each move the chip will move down one level until at the end of the game it will be at the bottom level. Suppose, in general, that for some $0 \le i < n$ the chip is on the i-th level at point H. First suppose that the quantification of x_{i+1} is existential. Then Spoiler moves the chip to the $i+1$-st level to some point H' such that (H, H') is an acceptable move. Otherwise, suppose that the quantification of x_{i+1} is universal. Then Verifier moves the chip to the $i+1$-st level to some point H' such that (H, H') is an acceptable move. At the end of the game Verifier wins if the position H reached on the bottom level is marked Yes. Spoiler wins if H is marked No.

Claim 7.1.6 *Verifier wins the Truth Game if and only if $A \in T_\alpha$*

We can call an intermediate point at level i winning if the game is a win for Verifier when it is started at that point. A simple backwards induction on i gives that H is a winning position if and only if

$$\mathrm{cl}_{t_i}(x_1,\ldots,x_i) \cong H \Rightarrow Q_{x_{i+1}} \cdots Q_{x_k} P(x_1,\ldots,x_k)$$

is a theorem of T_α.

This gives a more constructive proof of Theorem 7.1.5. When α is good as defined above the values t_i are constructive, as are the possible t_i-types H and the possible moves (H, H'). Further the vertices marked Yes on the bottom level are constructively defined. Hence whether $A \in T_\alpha$ can be determined by solving a finite perfect information game. The complexity of the determination, however, will depend on the size of the board which in turn depends on the values t_i. As the t_i may be arbitrarily large it is difficult to give a complexity result for general α. We return to this game giving a more definitive complexity condition with Theorem 9.2.6.

7.2 Countable Models

Gödel's Completeness Theorem gives the existence of countable models (i.e., graphs) satisfying the theory T_α. Here we shall give a specific countable model G_α . For convenience, label the vertices with the positive integers. The idea will be to satisfy all of the extension statements by creating witnesses [somewhat akin to the proof of Gödel's Theorem itself] in a minimal possible way.

7.2.1 Construction

By a *witness request* we mean a safe rooted graph (R, H) together with an ordered r-tuple (where R has size r) of distinct positive integers $n_1, \ldots, n_r$. (When $R = \emptyset$ we exclude the second part.) For example, let $\alpha = \pi/7$ and H consist of two roots with a common neighbor. We can interpret a typical witness request as: we need a y adjacent to 17 and 257. If H is a path of length two with one endpoint as the single root it could be: we need y_1, y_2 with y_1 adjacent to 28 and y_2 adjacent to y_1.

For each (R, H) there are a countable number of r-tuples. There are a countable number of (R, H). Fortunately, countability is a most robust notion and the number of witness requests is itself countable. We place these witness requests in a countable list.

We begin our (infinite) procedure to create G_α with no adjacencies between the vertices and every vertex considered unused. Once two vertices are made adjacent they remain that way forevermore. Once a vertex is considered used it remains that way forevermore. At every stage the used vertices will be an initial interval of the integers.

In a general position suppose we come to the next witness request with (R, H) and $n_1, \ldots, n_r$. First a technical point. If the n_i are not all used take another witness request all of whose vertices are already used and consider it first. We can always do this as, indeed, there are an infinite number of requests with $R = \emptyset$. [Indeed, the simplest requests are with H having v

vertices, no edges and no roots – requesting v new vertices.] After we've bumped the queue a finite number of times we'll get to what had been our next witness request. Hence all witness requests will eventually be considered. Here is how we "satisfy" a witness request. Let $m+1,\ldots,m+v$ be the first v unused vertices. Create an (R,H) extension of the m's over the n's, joining n_i to $m+j$ when x_i is adjacent to y_j in H and joining $m+i$ to $m+j$ when y_i is adjacent to y_j in H. Henceforth, consider $m+1,\ldots,m+v$ used. We call $B=[m+1,m+v]$ a *block* and say B is *over* $X=\{n_1,\ldots,n_r\}$. By an *initial segment* we mean the set I of used vertices at the end of some stage. Equivalently, $I=[1,m]$ for some m and is the union of blocks.

That gives the construction. Because the list of witness requests is countable at any stage only a finite interval of integers have been used and so unused integers are available to satisfy the request. The graph so constructed has a number of intriguing properties.

Claim 7.2.1 *Let $I=[1,m]$, $B=[m+1,m+v]$, $I'=I\cup B=[1,m+v]$ where B is a block and I,I' are initial segments. Then (I,I') is a safe extension.*

Proof: B is over $X\subseteq I$ so $(X,X\cup B)$ is a safe extension. There are no edges from B to $I-X$. Adding new roots $I-X$ gives (I,I'), which is safe by Property 4.1.10.

Claim 7.2.2 *Let $I=[1,m]$, $I'=[1,m']$ be initial segments with $m<m'$. Then (I,I') is a safe extension.*

Proof: Consider the tower $I=I_0\subset I_1\subset\ldots\subset I_l=I'$ of consecutive initial segments. Each (I_i,I_{i+1}) is safe by Claim 7.2.1 so (I,I') is safe by Property 4.1.14.

Claim 7.2.3 *Initial segments have no dense extensions.*

Proof: Let $I=[1,n]$ be an initial segment. Any extension (I,H) is a subextension of some (I,I') where I' is an initial segment. From Claim 7.2.2 (I,I') is safe so none of its subextensions can be dense.

For our purposes the following weaker form shall prove useful.

Claim 7.2.4 *Initial segments have no rigid extensions.*

Theorem 7.2.5. *The construction described above gives a countable model G_α for the theory T_α.*

First consider Generic Extension. Let (R,H) be a safe extension and let integers $n_1,\ldots,n_r$ be given. Let t be arbitrary. At some time a witness request for (R,H) with $n_1,\ldots,n_r$ would come up. Let $I=[1,m]$ be the initial segment at this stage and let block $B=[m+1,m+v]$ make the (R,H) extension. Suppose (H,H_1) were rigid. Adding new roots $I-V(H)$, by Property 4.1.12, $(I,H_1\cup I)$ is rigid or trivial in that $I=V(H_1)\cup I$.

From Claim 7.2.4 it cannot be rigid so we must have $V(H_1) \subseteq I$. From our construction block B can be adjacent in I only to the $n_1, \ldots, n_r$. Therefore block B is a t-generic (R, H) extension over $n_1, \ldots, n_r$.

We note that G_α actually has a property somewhat stronger than Generic Extension. For every safe extension (R, H) and vertices $n_1, \ldots, n_r$ we have found a block B. This B is a t-generic (R, H) extension over $n_1, \ldots, n_r$ *simultaneously* for all t.

The NonExistence schema is also simple. It suffices to show that no H exists where H is a minimal graph with $v(H) - \alpha e(H) < 0$. Consider the last time when the initial segment $I = [1, n]$ did not contain H. Then $H' = H|_I$ has v' vertices, e' edges with $v' - \alpha e' \geq 0$ (equality only if $v' = e' = 0$) and so (I, H) is dense. This contradicts Claim 7.2.3.

We conclude with further properties of G_α that shall prove useful in the next section.

Claim 7.2.6 *Let $I = [1, n]$ be an initial segment. Let $B = [m+1, m+v]$ with $m > n$ be a block over $X \subseteq I$. Let $J = [n+1, m]$. Then there are no edges between B and J.*

Proof: From the construction the only edges from B to $[1, m]$ are those to X given by the (R, H) extension.

Claim 7.2.7 *Let I, B, J, X be as in Claim 7.2.7. Then $(I \cup B, I \cup B \cup J)$ is a safe extension.*

Proof: $(I, I \cup J)$ is a safe extension by Claim 7.2.2. Adding new roots B leaves a safe extension by Claim 7.2.6 and Property 4.1.10.

7.2.2 Uniqueness of the Model

Theorem 7.2.8. *The G_α described above is unique. That is, if two orderings of the witness requests produce G_α, G'_α then these graphs are isomorphic.*

Proof: We create the isomorphism between the two graphs by a back and forth argument similar to the one applied in the proof of Theorem 0.1.5 for the p constant case. We'll create the isomorphism $\Psi\colon G_\alpha \to G'_\alpha$ in stages. At every stage Ψ will give a graph isomorphism between restrictions of G_α, G'_α to finite sets. These maps Ψ will be in one of the following three forms:

- $\Psi\colon I \to I'$ where I, I' are initial segments.
- $\Psi\colon I \cup B \to I'$ where $I' = [1, n']$ is an initial segment of G'_α, $I = [1, n]$ is an initial segment of G_α, $B = [m+1, m+v]$ is a block of G_α, $m > n$ and B is over $X \subseteq I$. We set $J = [n+1, m]$ in this case.

• $\Psi: I \to I' \cup B'$ where $I = [1, n]$ is an initial segment of G_α, $I' = [1, n']$ is an initial segment of G'_α, $B' = [m'+1, m'+v']$ is a block of G'_α, $m' > n'$ and B' is over $X' \subseteq I'$. We set $J' = [n'+1, m']$ in this case.

Suppose the second case. From Claim 7.2.7 $(I \cup B, I \cup B \cup J)$ is a safe extension, call it (R, H). At some stage in the construction of G'_α a witness request for an (R, H) extension over I' is received, with the ordering of the vertices I' determined by Ψ. At that stage a block B' is used, making $(I', I' \cup B')$ an (R, H) extension. We extend Ψ to $\Psi^+: I \cup B \cup J \to I' \cup B'$. Now $I \cup B \cup J$ is an initial segment. If $I' \cup B'$ is an initial segment we are in the first case, otherwise we are in the third case.

The third case is equivalent to the second case with the roles of G_α, G'_α reversed. From the third case we end in either the first case or the second case.

The first case is similar. Let B be the first block after I. Then B is over $X \subseteq I$ so that $(I, I \cup B)$ is a safe extension. At some stage in the construction of G'_α a witness request for an (R, H) extension over I' is received, with the ordering of the vertices I' determined by Ψ. At that stage a block B' is used, making $(I', I' \cup B')$ an (R, H) extension. We extend Ψ to $\Psi^+: I \cup B \to I' \cup B'$. Now $I \cup B$ is an initial segment. If $I' \cup B'$ is an initial segment we are in the first case, otherwise we are in the third case.

In all cases we have extended Ψ. Initially we may consider Ψ to have domain and range the empty set so that the first case applies. Let C be the first block of G_α not in the domain of Ψ at some stage. If we are in the first or second cases then C will be in the domain of Ψ^+. Two consecutive iterations (this is the "back and forth" aspect) cannot both be in the third case. Hence C must be in the domain of Ψ^{++}. Every vertex of G_α lies in some block and so will be in the domain of Ψ after a finite number of iterations. Similarly, every vertex of G'_α will be in the range of Ψ after a finite number of iterations. Hence the final Ψ at the end of this infinite process gives the desired isomorphism.

These graphs G_α are intriguing countable graphs deserving of further study. What, for example, can one say about the automorphism group of G_α?

7.2.3 NonUniqueness of the Model

Here we outline the argument that the theory T_α is *not* $\aleph_0$-categorical. That is, besides G_α there exist other (nonisomorphic) countable graphs which are also models for T_α.

Let us define the closure $\mathrm{cl}(X)$ as the union of the t-closures $\mathrm{cl}_t(X)$ for all t. Note that being in the closure is not a first order property. From Claim 7.2.4 $\mathrm{cl}(I) = I$ for all initial segments I. Every finite X has $X \subseteq I$ for some initial segment so that $\mathrm{cl}(X) \subseteq I$. In particular, finite sets have finite closures.

Our object will be to create a countable model G with an element c so that the closure of $\{c\}$ is infinite.

Consider graphs H with v vertices and e edges and a designated vertex x such that $(\emptyset, H)$ is a safe extension but $(\{x\}, H)$ is a rigid extension. [For this to occur a necessary condition is that $v - e\alpha > 0$ but $(v-1) - e\alpha < 0$. The actual construction of such H is a technical challenge that we omit.] The random graph will almost surely have (many) copies of H but most vertices will not lie in a copy of H.

The heart of the technical argument is constructing an infinite sequence H_i of such graphs such that, letting $H^{\leq i}$ denote the union of $H_1, \ldots, H_i$ where the designated vertices are identified, the extensions $(\emptyset, H^{\leq i})$ are all safe. [With H_i having v_i vertices and e_i edges this requires all $v_j - 1 - e_j\alpha < 0$ but all $1 + \sum_{j \leq i}(v_j - 1 - e_j\alpha) > 0$. From a number theory perspective this is possible since e_j, v_j can be found with $v_j - 1 - e_j\alpha$ negative but arbitrarily close to zero. The requirements of safe and rigid for the extensions do require further, omitted, work.] In any model G of T_α and for every i there will exist a vertex c contained (in the designated position) in a copy of $H^{\leq i}$. We will construct a model G of T_α with a particular vertex c contained (in the designated position) in a copy of $H^{\leq i}$ for all i. Such a c will have infinite closure and therefore the model G cannot be isomorphic to G_α where every singleton has finite closure.

Creating G is done by a method familiar to logicians. Add to the first order language a constant symbol c. Let A_i be the first order sentence that c lies in a copy of $H^{\leq i}$ in the designated position. Consider the theory T_α with the infinite sequence of sentences A_i appended, call it T_α^+. Now any finite subset of T_α^+ has a model. Indeed, consider the old T_α with $A_1, \ldots, A_i$ appended. (Any finite subset is contained in this for some i.) Any countable model G of T_α does have a copy of $H^{\leq i}$ and identifying c with the appropriate element of $H^{\leq i}$ makes G a model. Thus every finite subset of T_α^+ is consistent. (Reality must be consistent!) By the *compactness argument* T_α^+ itself is consistent. (If it were not consistent there would be a proof of f – false – but this proof would be finite and so use only a finite number of the axioms.) But then by the Gödel Completeness Theorem T_α^+ has a model G which has a special vertex c. As all A_i are satisfied for this c the closure of c is, as desired, infinite.

The above argument is made more precise and is extended greatly in the author's [18]. Instead of just H_i two graphs H_i^1, H_i^2 are given. For any choice H_i from the two a model is created with an element c in a copy of every H_i. A countable number of binary choices lead to a continuum of possible values for the closure of c and then to a continuum of nonisomorphic countable models of T_α

7.3 A Continuum of Complete Theories

One final note. For every rational number $a/b \in (0,1)$ there is a graph H such that the threshold function for the existence of H is $n^{-a/b}$. (This is a technical matter, creating a strictly balanced graph with v vertices e edges and $v/e = a/b$.) Any two irrational $\alpha, \alpha' \in (0,1)$ are separated by some such a/b and hence their theories $T_\alpha, T_{\alpha'}$ are different. This also means (as the theories are complete) that the models $G_\alpha, G_{\alpha'}$ are nonisomorphic. Thus we have a continuum of complete theories and a continuum of specific nonisomorphic countable graphs.

8. Near Rational Powers of n

Those with experience in random graphs should not be surprised that the Zero-One Law does not hold when $p = n^{-1/3}$. Take a strictly balanced H with e edges, v vertices and $v/e = 1/3$ – more specifically, take $H = K_7$. The number X of copies of K_7 has $E[X] = \binom{n}{7}p^{21} \to 1/7!$. From Janson's Inequality 5.0.4, or even the original methods of Erdős and Rényi, $\Pr[X = 0] \to e^{-1/7!}$. Here $X = 0$ is the first order $\neg\exists K_7$ and $e^{-1/7!}$ is certainly neither zero nor one. One might expect (as, indeed, this author once conjectured) that for any first order A $\lim_{n\to\infty} \Pr\big[G(n, n^{-1/3}) \models A\big]$ would exist and one might even hope for a description of the limiting probabilities as in Section 3.6. We shall see in this chapter that very much the opposite is true, that there are first order A for which $\Pr[A]$ behaves very strangely at and near $n^{-1/3}$.

The choice of $n^{-1/3}$ is a technical convenience. Similar results apply near $n^{-\alpha}$ for any rational $\alpha \in (0, 1)$.

8.1 Infinitely Many Ups and Downs

8.1.1 In the Second Order World

Here we give a first order sentence A that may be a surprise to those who study random graphs. Suppose $p = n^{-\alpha}$ and

$$\frac{1}{k+1} < \alpha < \frac{1}{k} \tag{11}$$

where we'll further assume, say, $k \geq 10$ to avoid trivialities. Our A will have the property that $\lim_{n\to\infty} \Pr\big[G(n,p) \models A\big]$ is one when k is even and zero when k is odd. Roughly speaking, A has an infinite number of threshold functions.

Let $S(x, y, z)$ be the set of w with $w \sim x \wedge w \sim y \wedge w \sim z$, the set of common neighbors of x, y, z. The maximum size $\big|S(x, y, z)\big|$ is the maximum t for which there exists in $G(n,p)$ the complete bipartite graph $K_{3,t}$. For fixed t this is covered by the classic work of Erdős and Rényi. $K_{3,t}$ is a strictly balanced graph with $v = 3 + t$ vertices and $e = 3t$ edges. The threshold function for the existence of a copy of $K_{3,t}$ is then $n^{-(3+t)/3t}$. For $p \gg n^{-1/3-1/t}$ there almost surely exists a copy while for $p \ll n^{-1/3-1/t}$ there almost surely does not exist a copy. Now we turn things around and let α be as above. Then almost surely there does exist a $K_{3,k}$ but does not exist a $K_{3,k+1}$. Thus the maximal $|S(x,y,z)|$ is k. We can take a dynamic viewpoint here, thinking of p approaching $n^{-1/3}$ from below or, with $p = n^{-\alpha}$, α approaching $1/3$ from above. Then $\max|S(x,y,z)|$ is running through the positive integers, increasing by one at each $\alpha = \frac{1}{3} + \frac{1}{k}$. We would now *like* to say

$$\text{the maximal } |S(x,y,z)| \text{ is even}$$

For convenience we'll say S is even when $|S|$ is even and S is bigger than T when we mean $|S| > |T|$. The word maximal can be replaced by the word bigger. If one could say

$$\forall_{x,y,z} \text{ if no } S(x',y',z') \text{ is bigger than } S(x,y,z) \text{ then } S \text{ is even}$$

then we would have our desired property. The two difficulties (which are handled quite similarly) are then in saying S is bigger than T and that S is even in the first order language.

The key will be to write a sentence in the second order language and then to give an appropriately equivalent statement in the first order language. In the second order language one is able to quantify over i-ary predicates. When $i = 1$ this is quantification over sets. For example, one could say that G is 3-colorable by saying there exist three unary predicates R, Y, G (Red, Yellow, Green) such that for all x precisely one of $R(x), Y(x), G(x)$ hold and for all $x \neq y$ with $x \sim y$ one doesn't have $R(x) \wedge R(y)$ nor $Y(x) \wedge Y(y)$ nor $G(x) \wedge G(y)$. Recall Theorem 3.3.3 that 3-colorability is not expressible in first order. This example shows that we cannot in general replace second order by first order.

We shall actually be interested in the case $i = 2$, quantification over binary predicates. First, consider saying that a set S is even. We say that a binary predicate $R(x,y)$ *witnesses* that S is even if

1. R is symmetric and areflexive
2. For all $x \in S$ there is a unique $y \in S$ with $R(x,y)$

When R witnesses that S is even the graph on S given by R is a 1-factor and so S must indeed be even. Suppose, as will always be the case, that membership in S is itself given by a first order predicate. Then S being even is given by the second order predicate that there exists R witnessing that S is even.

Similarly, suppose we want to say set S is bigger than set T. We say that a binary predicate $R(x,y)$ witnesses that S is bigger than T if

1. R is symmetric and areflexive
2. For all $x \in T - S$ there is a $y \in S - T$ with $R(x,y)$
3. If $x, x' \in T - S$ and $y \in S - T$ and $R(x,y)$ and $R(x',y)$ then $x = x'$
4. For some $y \in S - T$ there is no $x \in T - S$ with $R(x,y)$

Such an R would give a one-to-many relationship between all of $T-S$ and a proper subset of $S-T$ so that $S-T$ would be bigger than $T-S$ and hence S would be bigger than T. Suppose, as again will always be the case, that membership in S, T are given by first order predicates. Then membership in $S-T$ and $T-S$ are also given by first order predicates. Then S being bigger than T is given by the second order predicate that there exists R witnessing that S is bigger than T.

With this expanded language we can give the desired sentence A. We say there exist x, y, z such that for no x', y', z' does there exist R witnessing that $S(x', y', z')$ is bigger than $S(x, y, z)$ (as defined above) and such that (as also defined above) there is an R witnessing that $S(x, y, z)$ is of even size. The statement $w \in S(x, y, z)$ is replaced by the first order $w \sim x \wedge w \sim y \wedge w \sim z$ and similarly for $w \in S(x', y', z')$. The only problem is that we want to limit ourselves to first order sentences and the phrase "there exists R" is definitely second order.

8.1.2 Replacing Second Order by First Order

As above, we are examining $G(n,p)$ with $p = n^{-\alpha}$ where $\frac{1}{3}+\frac{1}{k+1} < \alpha < \frac{1}{3}+\frac{1}{k}$. We have first order defined sets S [the $S(x, y, z)$ above] and we want to translate the second order sentence that S is even [bigger will be similar] to an appropriately equivalent first order sentence. Note, critically, that the sets S we will be dealing with all have size at most k. If k were, say, 10 we could simply write a first order property that said S had size zero, two, four, six, eight or ten. The difficulty is to write a single sentence that works for all k.

For any set S and any vertex $v \notin S$ we let R_v be the binary relation on S defined by:

$$R_v(x, y) \colon x \neq y \wedge \exists_u (u \sim v \wedge u \sim x \wedge u \sim y)$$

Each $v \notin S$ defines a symmetric areflexive binary R_v on S. Roughly speaking [but this will not quite work] we will replace "There exists R which witnesses S being even" with "There exists v such that R_v witnesses S being even". The second is an entirely first order predicate. We know that if some R_v witnesses S being even then S is indeed even. The difficulty is the converse. When S is even there is a binary $R(x, y)$ witnessing that S is even – just pick any 1-factor for R. But is there an R of the form R_v? We'd like a representation theorem that says given an R on S there exists v with R_v the same as R on S. What we actually get is somewhat weaker.

Theorem 8.1.1 (Representation Theorem). *Let $k \geq 10$, B be arbitrary positive integers. Let $\frac{1}{3}+\frac{1}{k+1} < \alpha < \frac{1}{3}+\frac{1}{k}$ and let $G \sim G(n,p)$ with $p = n^{-\alpha}$. Let S have $r \leq B$ vertices and let R be a symmetric areflexive binary relation with $R(x, x')$ holding for t unordered pairs $x, x' \in S$* where t is at most $k/3$. *Almost surely for all such S, R there exists $z \notin S$ with R_z the same as R*

on S. That is, for all distinct $x, x' \in S$ there is a common neighbor for x, x', z exactly when $R(x, x')$.

Proof: Label the vertices of S by $x_1, \ldots, x_r$. Label the unordered pairs for which R holds $x_{a(i)}, x_{b(i)}$ for $1 \leq i \leq t$. Consider the rooted graph (S, H) with roots S and nonroots $y_1, \ldots, y_t, z$ where for each $1 \leq i \leq k$ we have $z, x_{a(i)}, x_{b(i)}$ adjacent to y_i and no other edges. (S, H) has $v = 1 + t$ nonroots and $e = 3t$ edges. We calculate

$$v - \alpha e < 1 + t - \left(\frac{1}{3} + \frac{1}{k}\right)(3t) = 1 - \frac{3t}{k}$$

The upper bound on t in our assumptions makes this positive so that (S, H) is a sparse extension, The subextensions of (S, H) are simple to describe and one sees that (S, H) is indeed a safe extension. By the Generic Extension Theorem 5.3.1 we have that almost surely for all $x_1, \ldots, x_r$ there exist $y_1, \ldots, y_t, z$ having precisely these edges and being a 1-generic extension. We claim R_z on S is precisely R. It has been designed so that $R(x, x') \Rightarrow R_z(x, x')$ but what about the converse. If distinct $x, x' \in S$ do not satisfy $R(x, x')$ can x, x', z have a common neighbor w? No! w couldn't be one of the y_j since the extension – by 0-genericity – has no additional edges. What if w was an outside vertex. If that did occur w would be in a rigid extension over z, x, x' (with one nonroot and three edges, as $1 - 3\alpha < 0$, using the fact that $\alpha > \frac{1}{3}$) and adjacent to z which would contradict $y_1, \ldots, y_t, z$ being 1-generic over $x_1, \ldots, x_r$.

Note that the bound B on the number of vertices plays no role – in examples we shall have $B = k$ or $B = 2k$ – but that the number of instances t of R is limited to $k/3$. To witness that a k-set S is even, however, we need a graph on $k/2$ edges. This has a simple technical fix. We call S *pseudoeven* if there exist v, w such that the union $R_v \cup R_w$ witnesses that S is even. When membership in S is given by a first order predicate S being pseudoeven is given by a first order predicate. When S is pseudoeven it must be even. As a partial converse (but precisely what we need), let S be any even set on at most k vertices. Take a 1-factor on S and split it into two graphs H_1, H_2, both with $\leq \lfloor k/3 \rfloor$ edges. Then there will be v, w with R_v, R_w being H_1, H_2 respectively on S. Then $R_v \cup R_w$ witnesses that S is even so S is pseudoeven. That is, almost surely pseudoeven is even on all sets of size at most k.

To witness that S is bigger than T when both have at most k vertices takes less than k edges on a set of size at most $2k$. These at most $k - 1$ edges can be split into four groups, each of size at most $k/3$. We call S *pseudobigger* than T if there exist v, w, u, s such that the union $R_v \cup R_w \cup R_u \cup R_s$ witnesses that S is bigger than T. As above, when S is pseudobigger than T it must be bigger and almost surely for every pair of sets S, T, each of size at most k, with S bigger than T, we do have S pseudobigger than T.

8.1.3 Are First Order Properties Natural?

Let's review what our sentence A looks like in the first order world. We write $R_v(a,b)$ as shorthand for

$$a \neq b \wedge \exists_z (v \sim z \wedge a \sim z \wedge b \sim z)$$

We write $w \in S(x,y,z)$ as shorthand for $w \sim x \wedge w \sim y \wedge w \sim z$. $w \notin S$ is shorthand for $\neg(w \in S)$. We write $S(x,y,z)$ even as shorthand for

$$\exists_{v,w} \forall_a \Big[a \in S(x,y,z) \Rightarrow \exists !_b \Big(b \in S(x,y,z) \wedge \big[R_v(a,b) \vee R_w(a,b) \big] \Big) \Big]$$

The fragment $S(x,y,z)$ is bigger than $S(x',y',z')$ has the form $\exists_{v,w,u,s} B$. Write

$$R(a,b) \text{ for } R_v(a,b) \vee R_w(a,b) \vee R_u(a,b) \vee R_s(a,b)$$

$$\text{Left}(a) \text{ for } a \in S(x,y,z) \wedge a \notin S(x',y',z')$$

$$\text{Right}(a) \text{ for } a \in S(x',y',z') \wedge a \notin S(x,y,z)$$

Then B can be written $C \wedge D \wedge E$ with

$$C \text{ for } \forall_a \Big[\text{Left}(a) \rightarrow \exists_b \big(\text{Right}(b) \wedge R(a,b) \big) \Big]$$

$$D \text{ for } \forall_{a,a',b} \Big[\big(\text{Left}(a) \wedge \text{Left}(a') \wedge \text{Right}(b) \wedge R(a,b) \wedge R(a',b) \big) \Rightarrow a = a' \Big]$$

$$E \text{ for } \exists_b \Big[\text{Right}(b) \wedge \neg \exists_a \big(\text{Left}(a) \wedge R(a,b) \big) \Big]$$

Finally, A is the sentence:

$$\forall_{x,y,z} \big[\neg \exists_{x',y',z'} S(x',y',z') \text{ bigger than} S(x,y,z) \big] \Rightarrow S(x,y,z) \text{ even}$$

All of these shorthands can certainly be unravelled in principle and one is left with a very long (but definitely finite!) first order sentence A which has an infinite number of ups and downs as described at the beginning of this section.

The feeling in the Random Graph community is that natural properties behave nicely with respect to the evolution of $G(n,p)$. They have clearly defined threshold functions and should only have a finite number of "ups and downs". But the word natural is not well defined. If we want to talk about all statements of a certain type then it seems one must tread into logical waters, one must define the language in which the statements are expressed. For the logicians the first order language is a natural choice. But how does it compare to the graph theorist's intuitive feel for the term natural. We have seen that certain properties that any graph theorist would term natural, such as connectivity, are not expressible in the first order world. Now we have seen an example in the other direction. The sentence A constructed above is in the first order world but few graph theorists would term it natural. The terms natural and first order are simply far apart.

8.2 Existence of Finite Models

This section does not examine random graphs. Rather, given a first order sentences A we ask whether there exists some finite graph G satisfying A. The result is in the negative.

Theorem 8.2.1. *There is no decision procedure with inputs first order sentences A that decides if there is some finite graph satisfying A.*

This is actually a case of a more general result, known as the Trakhtenbrot-Vaught Theorem. We instead deduce it from a classic result: the undecidability of the Post correspondence problem.

Let Σ be a finite alphabet and let a finite set of pairs u_i, v_i, $1 \le i \le s$, with $u_i, v_i \in \Sigma^*$ (that is, strings from alphabet Σ) be given. The Post correspondence problem is to determine whether there exists a word w which may be parsed as $x_1 \cdots x_l$ and $y_1 \ldots y_l$ for some $l > 0$ so that each pair x_j, y_j is one of the pairs u_i, v_i. The classic result of Emil Post is that there is no decision procedure to determine, given Σ and the u_i, v_i, whether such a w exists.

To prove Theorem 8.2.1 we find, given Σ and the u_i, v_i, an A which has a finite model G if and only if the Post correspondence problem has a solution. We begin by allowing some language from second order. For a set graph H on vertex set S to provide a model for the Post problem we first require

1. S has a linear order $<$.
2. S has unary predicates U_α for each $\alpha \in \Sigma$ so that for each $v \in S$ precisely one $U_\alpha(v)$ holds.
3. S has a binary relation $\textsc{Cut}(x, y)$.

Call x a top cut if some $\textsc{Cut}(x, y)$ and similarly call y a bottom cut. These shall represent the left ends of corresponding words. We require that when $\textsc{Cut}(x, y)$ and $\textsc{Cut}(x', y')$ then x, x' and y, y' are in the same order. We further require $\textsc{Cut}(1, 1)$ where 1 is the first element of S under $<$. For each left cut x let x^+ denote the largest element (under $<$ in S) less than all left cuts $x' > x$. If there are no such x' let x^+ be the largest element of S. Define y^+ similarly. Now S is split into intervals by the $[x, x^+]$ as well as by the $[y, y^+]$. We require that whenever $\textsc{Cut}(x, y)$ then for some $1 \le i \le s$ the interval $[x, x^+]$ is u_i and the interval $[y, y^+]$ is v_i. As these are finite intervals this can be expressed in first order. Our second order A is that there exist $<, U_\alpha, \textsc{Cut}(x, y)$ with these first order properties.

We remove the second order aspects by a general procedure. A sentence is called existential second order if all of the second order quantifiers are existential and all occur at the beginning of the sentence. Observe that our sentence A is of this form.

Claim 8.2.2 *Let A be an existential second order sentence with maximal arity k. Then A may be constructively transformed into a first order A^* such that*

1. *If A^* holds for some finite graph G on m elements then A holds for some subgraph $H = G|_S$ with S having n elements and $n \leq m = O(n^k)$.*
2. *If A holds for some finite graph H on an n element set S then A^* holds for some extension G on m elements with $n \leq m = O(n^k)$.*

In particular, A holds for some finite graph if and only if A^* holds for some finite graph.

Proof: The transformation is explicit. We begin A^* with $\exists_v$, v a new variable. All segments $\exists_x P$ are now replaced with $\exists_x (x \sim v \wedge P)$. For each k-ary R in A^* (we use only $k = 1, 2$ in Theorem 8.2.1) we replace $\exists_R$ (recall as A^* is existential second order this only occurs once at the beginning of A^*) by $\exists_{v_1,\ldots,v_k}$ using new variables each time. Now each instance $R(x_1, \ldots, x_k)$ is replaced by

$$\begin{gathered}\text{There exist unique } w_1, \ldots, w_k \\ \text{with } v_i \sim w_i \text{ for } 1 \leq i \leq k, \\ \text{all } w_i \sim w_j \\ \text{and } w_i \sim x_i \text{ for } 1 \leq i \leq k\end{gathered}$$

We use different variable $w_1, \ldots, w_k$ for each instance of each R. Finally we add that every vertex is either v, adjacent to v, one of the $v_1, \ldots, v_k$ corresponding to some R, or is a w_i where $w_1, \ldots, w_k$ have the above property. This creates a totally first order sentence A^*.

If A^* holds for G then A holds for $G|_S$ where S is the set of neighbors of v. Let S have size n. An l-ary relation symbol R with u instances will yield $(u + 1)l$ extra vertices, the $v_1, \ldots, v_l$ and the $w_1, \ldots, w_l$ for each instance. This would be all elements of G so that $n \leq m \leq n + O(n^k)$. Conversely suppose A holds for H on vertex set S of size n. Add a vertex v adjacent to precisely S. For each k-ary relation symbol R add new vertices $v_1, \ldots, v_k$. For each instance of R, say $R(z_1, \ldots, z_k)$ add new vertices $w_1, \ldots, w_k$, with all $w_i \sim w_j$ and all $z_i \sim w_i \sim v_i$. Add no other vertices nor edges. Now A will hold for this larger graph G. Note that by insisting the w_i form a clique we have assured that there will be no other k-tuples $z_1, \ldots, z_k$ for which there exist $w_1', \ldots, w_k'$ with $z_i \sim w_i' \sim v_i$ where the w_i' form a clique. The total size m of G satisfies $n \leq m \leq n + O(n^k)$.

This completes Claim 8.2.2 and hence Theorem 8.2.1.

8.3 NonSeparability and NonConvergence

Here we examine the random graph $G(n, p)$ with $p = n^{-1/3}$. Similar results can be obtained at other rational powers in $(0, 1)$ and also for p appropriately close to those rational powers. Our results will be in the negative, that there are first order sentences A for which $\Pr\left[G(n, n^{-1/3}) \models A\right]$ does not behave well.

8.3.1 Representing All Finite Graphs

Given G and given distinct vertices $x, y, z; u$ we define a graph $H[x, y, z; u]$ as follows. The set of vertices of $H[x, y, z; u]$ has been previously defined in Section 8.1.1 as $S(x, y, z)$, the set of common neighbors of x, y, z. The edge relation has been previously defined as R_u: $w_1, w_2 \in S(x, y, z)$ are adjacent if u, w_1, w_2 have a common neighbor. Membership and adjacency in H are first order defined in terms of the parameters $x, y, z; u$. Note H might have no vertices, this is not a pointless concept!

Theorem 8.3.1. *Let H_0 be any finite graph. Then almost surely there exist $x, y, z; u$ with $H(x, y, z; u) \cong H_0$. Further almost surely for all x, y, z which have precisely $|V(H_0)|$ common neighbors there exists u with $H(x, y, z; u) \cong H_0$.*

Let H_0 have v vertices and e edges. We outline the argument which has two parts. First we want the existence of x, y, z with precisely v neighbors. Second we want for all $w_1, \ldots, w_v$ and every H_0 on them a u with R_u giving precisely the adjacencies of H_0 on the w's.

When x, y, z are selected at random their number of common neighbors has a binomial distribution $B[n-3, p^3]$ which has mean $(n-3)p^3 \sim 1$ and is asymptotically a Poisson distribution of mean one. The probability that it has value v is then $\sim 1/(ev!)$. Take K disjoint triples where $K = K(n) \to \infty$ slowly. Each triple fails to have precisely v neighbors with probability $\sim 1 - \frac{1}{ev!}$. These events are asymptotically independent [which takes some technical work] and the chance that no triple has the right number of neighbors is $[1 - \frac{1}{ev!}]^K$ which tends to zero.

Now consider $w_1, \ldots, w_v$ and possible u. A triple w_i, w_j, u has asymptotic probability e^{-1} of having no common neighbor. Let us think [again avoiding some technical work] of those events as mutually independent over all w_i, w_j, u. Then for each u the relation R_u on the w's is a random graph with probability $1 - e^{-1}$. Setting $A := \binom{t}{2} - e$, the number of nonedges of H_0, each R_v equals a particular labelled copy of H_0 with probability $(1 - e^{-1})^e (e^{-1})^A$. Letting a be the number of automorphisms of H_0 there are $v!/a$ such labellings and so $\Pr[R_v \cong H_0] \sim \gamma$ with $\gamma = (v!/a)(1 - e^{-1})^e (e^{-1})^A$. Here γ is a positive constant. With our assumption of independence the R_u are independent and so the probability that no u has $R_u \cong H_0$ is $\sim (1 - \gamma)^{n-v}$. This is an exponentially small failure probability. Thus even though there are $\binom{n}{v}$ possible v-sets and 2^E with $E = \binom{v}{2}$ possible H_0 (large, but constant) this is only polynomially many so that almost surely for every $w_1, \ldots, w_v$ the R_u range over all possible graphs.

While the argument above was made for r fixed it can, with somewhat more technical difficulty, be made to hold for all $r \leq r(n)$ where $r(n)$ is selected to go to infinity appropriately slowly. We state such a result, definitely not best possible, without proof.

Theorem 8.3.2. *Almost surely for every H_0 with less than* $\ln\ln\ln n$ *vertices there exist $x, y, z; u$ with $H[x, y, z; u] \cong H_0$. Further, for every x, y, z with $|V(H_0)|$ neighbors there is a u with $H[x, y, z; u] \cong H_0$.*

8.3.2 NonSeparability

Much work in random graph theory consists of calculating the limit probabilities of various events A in the random graph $G(n, p)$ for various $p = p(n)$. We restrict ourselves to first order properties A. When $p = n^{-\alpha}$ and $\alpha \in (0, 1)$ is irrational the limit probabilities are either zero or one and the recursive calculation of them is given by Theorem 7.1.5. Here we look at $p = n^{-1/3}$. The limit probabilities need not be zero nor one and worse, as we'll see in the next subsection, they need not even exist. Let us then just ask for a method of computing the limit when it is zero or one, the good cases. We call a Turing Machine a Separator if it takes as input first order sentences A and

1. If $\lim_{n\to\infty} \Pr\left[G(n, n^{-1/3}) \models A\right] = 1$ it outputs One.
2. If $\lim_{n\to\infty} \Pr\left[G(n, n^{-1/3}) \models A\right] = 0$ it outputs Zero

For other A – when the limiting probability either doesn't exist or lies in $(0, 1)$ we allow any response, the Turing Machine may output One or Zero or it may never halt.

Theorem 8.3.3. *There does not exist a Turing Machine Separator.*

Our strategy will be to reduce to Theorem 8.2.1.

For any first order sentence A on graphs let A^* be the sentence

$$A^*\colon \exists_{x,y,z,u} H[x, y, z; u] \models A$$

We can write A^* itself as a first order sentence. Any part $\exists_v$ in A is changed to $\exists_{v\in S(x,y,z)}$ which is written $\exists_v[v \sim x \wedge v \sim y \wedge v \sim z] \wedge \cdots$. Any part $w1 \sim w2$ is replaced by $w1 \neq w2 \wedge \exists_t t \sim w1 \wedge t \sim w2 \wedge t \sim u$. For example, $\forall_{w1}\exists_{w2} w1 \sim w2$ becomes

$$\exists_{x,y,z,u}\forall_{w1}\Big[\big[w1 \sim x \wedge w1 \sim y \wedge w1 \sim z\big] \Rightarrow \exists_{w2}$$

$$\big[w2 \sim x \wedge w2 \sim y \wedge w2 \sim z \wedge w1 \neq w2 \wedge \exists_t(t \sim w1 \wedge t \sim w2 \wedge t \sim u)\big]\Big]$$

If A holds for no finite graph then A^* will hold for no finite graph and so $\lim_{n\to\infty} \Pr\left[G(n, n^{-1/3}\right] \models A^*] = 0$. If A holds for some finite graph H_0, no matter how big, then almost surely in $G(n, n^{-1/3})$ there will be $x, y, z; u$ with $H[x, y, z; u] \cong H_0$ and so A^* would hold.

Suppose there did exist a Turing Machine Separator. For any first order A apply it to A^*. The answer would tell you if A held for some finite graph. This contradicts Theorem 8.2.1.

8.3.3 Arithmetization

A key element of Section 8.1 was the use of relational symbols to force a set S to be even. In that case we had a symmetric areflexive binary R with the property (on S) that $\forall_x \exists !_y R(x,y)$. In general, given a theory T with a possible variety of relational symbols the *spectrum* of T, denoted $Sp(T)$ is the set of nonnegative integers m such that there exists a model of T of size m. Thus in the above example the spectrum is the even integers. The study of spectra is a fascinating one, closely linked to problems in computational complexity. Here, however, we have a specific goal. We shall

1. Find an existential second order sentence whose spectrum has long gaps.
2. Use Claim 8.2.2 to give a first order sentence whose spectrum has long gaps.
3. Apply this in Section 8.3.4 to find a first order sentence whose limiting probability when $p = n^{-1/3}$ does not exist.

Our existential second order sentence begins

There exist binary relations
$x < y$, $D(x,y)$, $E(x,y)$, $T(x,y)$
such that

We require first that $<$ is an order:

$x < y \wedge y < z \Rightarrow x < z$
and for all x, y precisely one of
$x < y$, $y < x$, $x = y$ hold.

Any model with n elements can then be uniquely labelled $1, \ldots, n$. Now let 1 be shorthand for that element x with no $y < x$ and 2 for that x with precisely one $y < x$. We write $x + 1$ for that y (if it exists) with $x < y$ and no z satisfying $x < z < y$ and $x + 2$ for that y (if it exists) with $x < y$ and precisely one z satisfying $x < z < y$. Now we require

$D(1, y)$ if and only if $y = 2$
For $x \neq 1$, $D(x,y)$ if and only if
there exist z, w with $x = z + 1$, $y = w + 2$ and $D(z,w)$

In any model labelled $1, \ldots, n$ as above, $D(x,y)$ would hold if and only if $y = 2x$. Similarly require

$E(1, y)$ if and only if $y = 2$
For $x \neq 1$, $E(x,y)$ if and only if
there exist z, w with $x = z + 1$, $D(w,y)$ and $E(z,w)$

This forces $E(x,y)$ to hold if and only if $y = 2^x$. Similarly require

$T(1, y)$ if and only if $y = 2$
For $x \neq 1$, $T(x,y)$ if and only if
there exist z, w with $x = z + 1$, $E(w,y)$ and $T(z,w)$

This forces $E(x,y)$ to hold if and only if y is the tower function $y = T(x)$ as given by Definition 2.2. We define ARITH to be the second order existential sentence that there exist $<, D, E, T$ with these properties. Note that ARITH holds for all finite sets.

We now define the unary predicate LOGSTAR(x) so that it holds in our labelled model precisely when $x = \log^* n$ as given by Definition 2.2. Explicitly:

$$\text{LogStar}(x)\colon \left[(x' < x \Rightarrow \exists_y T(x,y)\right] \wedge \left[\neg\exists_{y,z}\left(T(x,y) \wedge y < z\right)\right]$$

With our particular goal in mind let us write a unary predicate MODC(z) representing that z modulo 100 is one of $1, \ldots, 50$. This is simply

$$\begin{gathered}\text{ModC}(1) \wedge \cdots \wedge \text{ModC}(50) \wedge \\ \neg\,\text{ModC}(51) \wedge \cdots \wedge \neg\,\text{ModC}(100) \\ \wedge \forall_{x,y} y = x + 100 \rightarrow \left[\text{ModC}(x) \Leftrightarrow \text{ModC}(y)\right]\end{gathered}$$

where the specific elements $1, \ldots, 100$ and the binary $y = x + 100$ can be written out in full. We define a second order existential sentence BIGGAP by:

There exist $<, D, E, T$ as above and x with LOGSTAR(x) and MODC(x)

Let S be a set of size s. Then BIGGAP holds on S if and only if $\log^* s$ modulo 100 is one of $1, \ldots, 50$.

We apply Claim 8.2.2 to ARITH give a first order sentence ARITH* on graphs. We further apply Claim 8.2.2 to give a first order sentence BIGGAP* on graphs. In our particular case we used maximal arity two.

Claim 8.3.4 *If* $\log^* t$ *modulo* 100 *is one of* $2, \ldots, 50$ *modulo* 100 *and* ARITH* *holds for a graph on* t *vertices then* BIGGAP* *holds for the same graph.*

Proof: We apply the same $v, <, D, E, T$. We have a set of size s with $\Omega(\sqrt{t}) \leq s \leq t$ on which the arithmetization takes place. But then $\log^* s$ modulo 100 is one of $1, \ldots, 50$ so that BIGGAP* also holds.

Claim 8.3.5 *If* $\log^* t$ *modulo* 100 *is one of* $52, \ldots, 99$ *modulo* 100 *then* BIGGAP* *does not hold on any graph on* t *elements.*

Proof: If it did there would be a set of size s with $\Omega(\sqrt{t}) \leq s \leq t$ on which the arithmetization takes place. But then $\log^* s$ modulo 100 is one of $51, \ldots, 99$ so that BIGGAP* does not hold.

8.3.4 NonConvergence

We fix $p(n) = n^{-1/3}$. Our object here is to give a first order sentence A such that $\lim_{n\to\infty} \Pr\left[G(n,p(n)) \models A\right]$ does not exist. We follow the ideas of Section 8.3.1.

For $x, y, z \in G$ we let $S(x,y,z)$ denote the set of their common neighbors. For $x, y, z, w \in G$ we define $H[x,y,z;w]$ as before: its vertex set is $S(x,y,z)$

and $a, b \in S(x,y,z)$ are adjacent if they share a common neighbor with w. We write $\text{ARITH}^*[x,y,z;w]$ and $\text{BIGGAP}^*[x,y,z;w]$ when the graph $H[x,y,z;w]$ has the first order properties ARITH^* and BIGGAP^* respectively. Note these are first order predicates in the variables x, y, z, w.

We call a set S' pseudobigger than a set S if there is a u such that R_u witnesses that S' is bigger than S in the sense of Section 8.1.2. That is, for all $a \in S - S'$ there exists a $b \in S' - S$ with u, a, b having a common neighbor but if $a, a' \in S - S'$ and $b \in S' - S$ we cannot have u, a, b having a common neighbor and u, a', b having a common neighbor unless $a = a'$. From Theorem 8.3.2, when both S' and S have size at most $\ln \ln \ln n$ and S' is bigger than S then S' is pseudobigger than S. [Technically, we are using the theorem up to sets of size $2 \ln \ln \ln n$. Actually it holds up to a power of $\ln n$.]

Now our desired sentence A is defined by saying there exist x, y, z, w such that

1. ARITH^* holds for $H[x,y,z;w]$
2. Whenever ARITH^* holds for $H[x',y',z';w']$, $S(x',y',z')$ is not pseudobigger than $S(x,y,z)$
3. BIGGAP^* holds for $H[x,y,z;w]$

Claim 8.3.6 *The above A has no limiting probability. That is,*

$$\lim_{n \to \infty} \Pr\left[G(n, n^{-1/3}) \models A\right]$$

does not exist.

Proof: First restrict to an infinite sequence of n for which $\log^* n$ is 99 modulo 100. There will be x', y', z', u' with $\left|S(x,y,z)\right| = t'$ for $t' = \lfloor \ln \ln \ln n \rfloor$ such that ARITH^* holds for $H[x',y',z';u']$. If x, y, z have $\left|S(x,y,z)\right| = t < t'$ then $S(x',y',z')$ will be pseudobigger than $S(x,y,z)$ as pseudobigger is the same as bigger on these small sets. Thus for x, y, z, u to have the first two properties above x, y, z must have $t > t'$ elements. But then $\log^* t$ would be one of $99, 98, 97, 96, 95$ modulo 100. By Claim 8.3.5 these x, y, z, u could not satisfy BIGGAP^*. Hence A has limiting probability zero along this subsequence.

Now restrict to an infinite sequence of n for which $\log^* n$ is 49 modulo 100. Take the biggest set $S = S(x,y,z)$ for which there exists u such that ARITH^* holds for x, y, z, u. Such a set exists with size $t = \Omega(\sqrt{\ln \ln \ln n})$. As pseudobigger implies bigger such x, y, z, u satisfy the first two properties. But now $\log^* t$ modulo 100 is one of $45, \ldots, 49$. By Claim 8.3.4 BIGGAP^* will hold for this $x, y, z; u$. Hence A has limiting probability one along this subsequence.

We have found a first order A for which $\Pr[A]$ does not converge as $n \to \infty$, indeed it has subsequences approaching both zero and one. Once again, the sentence is not one that the graphtheorist would consider natural!

8.4 The Last Threshold

We have seen that $\Pr[G(n,p) \models A]$ can exhibit strange behavior both as p approaches $n^{-1/3}$ from below (infinitely many ups and downs) and at $n^{-1/3}$ (nonconvergence). It may therefore come as a relief that the behavior for p slightly bigger than $n^{-1/3}$ is quite well behaved.

8.4.1 Just Past $n^{-\alpha}$: The Theory T_α^-

In this section we think of α as a *rational* number between zero and one. Our example is always $\alpha = 1/3$. The theory T_α^- shall be, in a sense to be described later, the limit of the theories T_β as β approaches α from below. It shall also be the almost sure theory for $p = p(n)$ appropriately larger than $n^{-\alpha}$ ($p > n^{-\alpha} \ln^K n$ for appropriate K will do here though this can be made much more precise) yet smaller than $n^{-\alpha+\epsilon}$ for any fixed positive ϵ.

We mimic the axiomatization of T_α given in Section 7.1.1 for irrational α. The axiomatization in turn built on the labelling of rooted graphs as dense or sparse given in Section 4.1 The strict dichotomy between dense and sparse now fails for α rational as we may well have $v - e\alpha = 0$. (In particular, for $\alpha = 1/3$, roots x_1, x_2, x_3 having a common neighbor nonroot y.) In our mind's eye we think of α is really being infinitesimally less than its rational value – we want a theory that works for $p = n^{-\alpha} n^{\epsilon(n)}$ where $\epsilon(n)$ is positive and approaching zero – such as $p = n^{-1/3} \ln^5 n$.

Let (R, H) be a rooted graph of type (v, e). We now call (R, H) sparse if $v - e\alpha$ is positive *or zero*. We call (R, H) dense if $v - e\alpha$ is negative. We have, by fiat, created a strict dichotomy – breaking all "ties" in favor of sparseness. We define safe and rigid in terms of sparse and dense as before. The t-closure $\mathrm{cl}_t(X)$ and the notion of $\mathbf{y}$ being a t-generic (R, H) extension of $\mathbf{x}$ are defined as before. We parrot the axiomatization of section 7.1.1.
Axiom Schema for T_α^-:

NonExistence Schema [For every H with v vertices e edges and $v - e\alpha < 0$] There does not exist a copy of H.

Generic Extension Schema [For every safe (R, H) and every $t \geq 0$] For all $\mathbf{x} = (x_1, \ldots, x_r)$ there exists $\mathbf{y} = (y_1, \ldots, y_v)$ forming a t-generic (R, H) extension over $\mathbf{x}$.

As in section 7.1 we want to show T_α^- is complete. We require, critically, the Finite Closure Theorem 4.3.2. We basically copy that proof. Set

$$\beta = \min \frac{e\alpha - v}{v}$$

the minimum over pairs of integers v, e with $v \leq t$ and $v - e\alpha < 0$. Set $K = \lceil \frac{r}{\beta} \rceil$. If there were a $\mathrm{cl}_t(R)$ of size at least K with $|R| = r$ there would be a sequence $R = S_0 \subset S_1 \ldots \subset S_u \subseteq \mathrm{cl}_t(R)$ with each (S_i, S_{i+1}) rigid of type (v_i, e_i) with $v_i \le t$ and $K \le \sum_{i=0}^{u-1} v_i \le K + t$. Since (S_i, S_{i+1}) is rigid we must have $v_i - e_i\alpha$ negative (*not zero* by our dichotomy rule) and so $v_i - e_i\alpha \le -\beta v_i$. Then the graph on S_u would have V vertices and E edges with $V - \alpha E < 0$ (as in the proof of Theorem 4.3.2) but this violates the NonExistence Schema.

(The asymmetry between p just below and just above $n^{-1/3}$ – or other rational powers – is one of the most intriguing aspects of the entire theory. The perspicacious reader may wonder where the symmetry breaks down. After all, we could define another notion of sparse and dense, breaking all ties in favor of denseness, calling (R, H) of type (v, e) dense if $v - e\alpha$ is negative or zero. We could then define t-closure and t-Generic extension and a theory T_α^+. It is precisely at the Finite Closure theorem that the symmetry breaks down. A rigid (S_i, S_{i+1}) could have $v_i - e_i\alpha = 0$. We therefore cannot limit the size of the t-closure. The theory T_α^+ is well defined above but it is not a complete theory. We can think of the example of section 8.1 as taking advantage of that. At $\alpha = 1/3$ the 1-closure of x_1, x_2, x_3 (in T_α^+) would contain all common neighbors of x_1, x_2, x_3 and there is no bound K to that number. The sets $S(x_1, x_2, x_3)$ can be of arbitrary size and they are central to creating a first order A with infinitely many ups and downs.)

The remainder of the proof of completeness for T_α^- has no surprises, it follows the argument for T_α, α irrational, given in Section 7.1 The central statement, shown by induction on the length, remains the same: For every predicate $P(\mathbf{x})$ there is an integer t such that, in T_α^-, $P(\mathbf{x})$ is logically equivalent to a disjunction of statements $\mathrm{cl}_t(\mathbf{x}) \cong H$. As $\mathrm{cl}_t(\emptyset) = \emptyset$ is in T_α^- every sentence is equivalent to either "true" or "false".

Now we can give a nice description of T_α^- as a "limit". Formally, define T_α^* as the set of first order A for which there exists a positive ϵ such that $A \in T_\beta$ for all (irrational, of course) β with $\alpha > \beta > \alpha - \epsilon$.

Theorem 8.4.1. $T_\alpha^- = T_\alpha^*$

Consider the axioms of T_α^-. First, NonExistence. If H has v vertices and e edges and $v - e\alpha < 0$ then there is a positive ϵ such that $v - e\beta < 0$ for every $\beta \in (\alpha - \epsilon, \alpha)$ and thus the nonexistence of H is in T_β for all such β and thus in T_α^*. Now, Generic Extension. Let (R, H) be safe, as defined for T_α^- and $t \ge 0$. For any $\beta < \alpha$ (R, H) will still be safe in T_β. An extension (H, H_1) of type (v, e) with $v \le t$ is dense in T_α^- if $v - e\alpha < 0$ and dense in T_β if $v - e\beta < 0$. When $v - e\alpha < 0$ there is a positive ϵ such that $v - e\beta < 0$ for all $\beta \in (\alpha - \epsilon, \alpha)$. There are only finitely many (v, e) with $v \le t$, pick the minimal ϵ of that finite set. Then for $\beta \in (\alpha - \epsilon, \alpha)$ the notion of t-generic is the same as that in T_α^-. The Generic Extension axiom for $(R, H), t$ in T_β thus is the same as the axiom in T_α^- and so the axiom in T_α^- is in T_β – and hence in T_α^*.

Any theorem A of T_α^- follows from a *finite* number of axioms $A_1, \ldots, A_u$ of T_α^-, for each there is a positive ϵ_i such that $A_i \in T_\beta$ for all $\beta \in (\alpha - \epsilon_i, \alpha)$.

Take $\epsilon = \min \epsilon_i$. For $\beta \in (\alpha - \epsilon, \alpha)$ all $A_i \in T_\beta$ so $A \in T_\beta$. Hence $A \in T^*_\alpha$. Thus $T^-_\alpha \subseteq T^*_\alpha$.

We have already seen, however, that T^-_α is a complete theory. There are only two possible extensions of a complete theory – either the (trivial) extension which is the theory itself or the (equally trivial) extension which is the *inconsistent* theory that contains all sentences. But we claim T^*_α cannot be inconsistent. If there was an inconsistency (a proof of "false") it would be derivable from a finite number of sentences $A_1, \ldots, A_u$. For each there would be a positive ϵ_i with $A_i \in T_\beta$ for all $\beta \in (\alpha - \epsilon_i, \alpha)$. Take $\epsilon = \min \epsilon_i$ and any irrational $\beta \in (\alpha - \epsilon, \alpha)$. Then all $A_1, \ldots, A_u \in T_\beta$ which would mean that T_β itself would be inconsistent, which it isn't. Thus T^*_α must be precisely T^-_α

The theory T^-_α shall also appear in Section 9.2.5. For now we note an important consequence of its completeness. Let A be any sentence. Either $A \in T^-_\alpha$ or $\neg A \in T^-_\alpha$, suppose the former. Then there is a positive ϵ such that $A \in T_\beta$ for all $\beta \in (\alpha - \epsilon, \alpha)$. In the later case the same holds for $\neg A$. In either case the infinitely many ups and downs that we've seen occur as $\beta \to \frac{1}{3}$ from above cannot occur when $\beta \to \frac{1}{3}$ from below. Moreover, using the results of the upcoming section, for any A there is a positive ϵ so that either A or $\neg A$ will hold almost always for *all* p with $n^{-\alpha} \ln^K n < p < n^{-\alpha+\epsilon}$

8.4.2 Just Past $n^{-\alpha}$: A Zero-One Law

Practitioners of random graph theory are well used to threshold functions that are products of powers of n and powers of $\ln n$. We saw in Section 3.6.4 the most celebrated case of this, connectivity having threshold function $\frac{\ln n}{n}$. In the first order world the pertinent sentence there was "Every vertex has a neighbor". We can give a similar statement near $p = n^{-1/3}$.

When does *every* triple of vertices have a common neighbor? At $p = n^{-1/3}$ a given triple of vertices may or may not have a common neighbors. Now suppose p gets bigger, parametrize $p = n^{-1/3}\lambda$. A triple has no common neighbor with probability $(1-p^3)^{n-3}$ which is roughly $e^{-p^3 n} \sim e^{-\lambda^3}$. We're led to parametrize $\lambda = c \ln^{1/3} n$ so that the probability becomes roughly n^{-c^3}. There are $\Theta(n^3)$ triples. When $c^3 > 3$ the failure probability (a triple not having a common neighbor) falls to $o(n^{-3})$ and so almost surely there is no failure, every triple has a common neighbor. While this argument gives only one side this extension property has been well studied and it is known that $n^{-1/3} \ln^{1/3} n$ is its threshold function. The (perhaps) surprise is that this is basically the last threshold function with lead term $n^{-1/3}$ – that, for example, $n^{-1/3} \ln^5 n$ is *not* a threshold function for any first order sentence.

We work in reverse direction from our usual approach. The complete theory T^-_α has been already given explicitly. For which $p = p(n)$ will all of the axioms of T^-_α hold almost surely? For those $p(n)$ we will have a Zero-One Law.

The NonExistence schema bounds $p(n)$ from above. Consider an H with v vertices, e edges and $v - e\alpha < 0$. Suppose $p = n^{-\alpha+o(1)}$. Then the expected number of copies of H is $\Theta(n^v p^e) = o(1)$. Let H be strictly balanced with v vertices, e edges and $v/e < \alpha$ so that $v - e\alpha < 0$. Those existence of such H has a threshold function $n^{-v/e}$. We must have $p < n^{-v/e}$ for the nonexistence of H to hold almost surely. Such H exist with v/e arbitrarily close to α. (Of course, there are rational numbers arbitrarily close to α and less than α – what we are using here is a technical result that for any rational number in $(0,1)$ there exists a strictly balanced graph with that number as its ratio of vertices to edges.) Thus we must have $p < n^{-\alpha+\epsilon}$ for all $\epsilon > 0$ – i.e., $p = n^{-\alpha+o(1)}$.

The t-Generic extension schema provide a technical challenge for the random graphtheorist, we content ourselves here with a summary of the results for $\alpha = 1/3$. The general t can be replaced by $t = 0$ – when p is sufficiently large that every $\mathbf{x}$ has an (R,H) extension $\mathbf{y}$ it will be the case that every $\mathbf{x}$ has a t-generic (R,H)-extension $\mathbf{y}$. When (R,H) has type $(v,3v)$ and is appropriately indecomposible the threshold is when $n^v p^{3v}$ reaches $\ln n$ which is $\Theta(n^{-1/3}\ln^{1/3v} n)$. This is maximized at $v = 1$ with the $n^{-1/3}\ln^{1/3} n$ threshold described above. To be even more precise – the threshold for every triple having a thousand common neighbors is slightly higher than that of every triple having a common neighbor. The exact result is: Let $p = n^{-1/3+o(1)}$ be such that for all u almost surely every triple has (at least) u common neighbors. Then $T^-_{1/3}$ is the almost sure theory for p and the Zero-One Law holds for p. In probabilistic language, the almost sure theory for $p = p(n)$ is T^-_α precisely when $p = n^{-1/3+o(1)}$ and we can write

$$p = n^{-1/3}\left[3\ln n + \omega(n)\ln\ln n\right]^{1/3}$$

where $\omega(n) \to \infty$.

Part III

Extras

9. A Dynamic View

Thus far we have employed what might be considered a static view of random graphs. We fix an edge probability $p = p(n)$ and look at the behavior all first order sentences A in $G \sim G(n,p)$. Erdős and Rényi, as discussed in Section 1.1.1, always thought of the random graph as *evolving* from empty to full. Now we shall fix an arbitrary first order sentence A and consider what happens to $\Pr[A]$ as the random graphs evolves.

9.1 More Zero-One Laws

9.1.1 Near Irrational Powers

We first look at p near $n^{-\alpha}$ for an irrational $\alpha \in (0,1)$. Recall from Chapter 7 that T_α represents the almost sure theory for $G(n,p)$ with $p = n^{-\alpha}$.

Theorem 9.1.1. *Let $\alpha \in (0,1)$ be irrational and let $A \in T_\alpha$. Then there exists $\epsilon > 0$ so that A holds almost surely in $G(n, p(n))$ for any function $p(n)$ satisfying $n^{-\alpha-\epsilon} < p(n) < n^{-\alpha+\epsilon}$ for all sufficiently large n.*

Proof: As $A \in T_\alpha$, A follows from some finite number of axioms of T_α and hence it suffices to show the existence of ϵ for each axiom of T_α. Any axiom of the NonExistence Schema has threshold function a negative rational power of n. An axiom of the Generic Extension schema follows from Theorem 5.3.1 which also holds in some neighborhood $n^{-\alpha-\epsilon} < p(n) < n^{-\alpha+\epsilon}$.

Note that the ϵ of Theorem 9.1.1 depends strongly on the choice of A. This must be the case. For any fixed positive ϵ there will be an irrational $\beta \neq \alpha$ with $|\beta - \alpha| < \epsilon$. The theories T_α, T_β are different so there will be an $A \in T_\alpha$ that has limit probability zero for $p = n^{-\beta}$.

Theorem 9.1.2. *If $p(n) = n^{-\alpha+o(1)}$ with $\alpha \in (0,1)$ irrational then $p(n)$ satisfies the Zero One Law.*

Proof: Immediate from Theorem 9.1.1.

9.1.2 Dense Random Graphs

Now we examine "large" $p = p(n)$.

Theorem 9.1.3. *Let A have quantifier depth $k+1$. Assume $p = p(n)$ satisfies*

$$p(n) \gg n^{-1/k} \ln^{1/k} n$$

and

$$1 - p(n) \gg n^{-1/k} \ln^{1/k} n$$

Then A holds either almost surely or almost never. Further, the choice of almost surely or almost never is identical for all such $p(n)$.

Proof: From Theorem 2.2.3 it suffices to show that $G(n, p(n))$ almost surely has the k-Alice's Restaurant property that for every k vertices $x_1, \ldots, x_k$ there are witnesses z having every possible adjacency pattern to the xs. Suppose $p \le \frac{1}{2}$. Let ϵ be the probability that z is adjacent to $x_1, \ldots, x_i$ and nonadjacent to $x_{i+1}, \ldots, x_k$. Then $\epsilon = p^i (1-p)^{k-i} \ge p^k \gg n^{-1} \ln n$. The probability that there is no witness z is $(1-\epsilon)^{n-k} \sim e^{-n\epsilon} \ll n^{-k}$. The case $p \ge \frac{1}{2}$ is similar. Hence almost surely there is a witness for all $O(n^k)$ choices for $x_1, \ldots, x_k$ and the adjacency pattern.

Theorem 9.1.4. *If $p(n) \gg n^{-\epsilon}$ and $1 - p(n) \gg n^{-\epsilon}$ for all positive ϵ then $p(n)$ satisfies the Zero One Law. Further, the complete theories for all such $p(n)$ are identical.*

This gives a natural extension of the fundamental Zero One Law Theorem 0.1.2, which may be regarded as the case $p = \frac{1}{2}$. In particular, note that all constant $p(n) = p$, $p \in (0, 1)$, are included in Theorem 9.1.4.

9.2 The Limit Function

We now aim for a dynamic view of $\Pr[A]$ as $p(n) = n^{-\alpha}$ evolves.

9.2.1 Definition

Definition 9.1 *Let A be a first order sentence. We define the* limit function *of A, denoted by $f_A(\alpha)$, by*

$$f_A(\alpha) = \lim_{n\to\infty} \Pr\big[G(n, n^{-\alpha}) \models A\big]$$

Equivalently,

$$f_A(\alpha) = \begin{cases} 1 & \text{if } A \in T_\alpha \\ 0 & \text{if } A \notin T_\alpha \end{cases}$$

The domain of f_A is the set of irrational $\alpha \in (0, 1)$.

It is critical that the domain of f_A be restricted to irrational α, as the results of Section 8.3.4 show that otherwise it might not be defined at all. Our Main Theorem 1.4.1 gives that $f_A(\alpha)$ is always well defined and in fact is always zero or one. We call $\alpha \in (0,1)$ (rational or irrational) a point of continuity for f_A if there is a positive ϵ so that f_A is constant (either zero or one) on $[\alpha-\epsilon, \alpha+\epsilon]$ where it is defined. Otherwise we call α a point of discontinuity. Theorem 9.1.1 gives that all irrational α are points of continuity, so that the set of points of discontinuity is a set of rational numbers. Theorem 9.1.3 gives that f_A is constant on $\left(0, \frac{1}{k-1}\right)$. Now we reexamine the proof of Theorem 1.4.1 and give further conditions on the set of points of discontinuity of the limit function f_A.

Our necessary conditions on f_A are best summarized by Theorems 9.2.5 and 9.2.6. These results are due to Gábor Tardos and the author [21]. It is believed that these conditions are sufficient – that for any f satisfying these conditions there is an A with $f = f_A$. Indeed, this result has been claimed but the technical challenges are daunting and a proof has not yet appeared in print.

9.2.2 Look-Ahead Functions

We fix k to be the quantifier rank of A. For convenience of notation we omit the dependence on k in the following definition. The look-ahead functions are designed to echo the strategy of Section 6.1 that allows Duplicator to almost always win the k-move Ehrenfeucht game with $G \sim G(n, n^{-\alpha})$.

Definition 9.2 *The* look-ahead functions $t_0(\alpha), \ldots, t_k(\alpha)$ *are defined by* $t_0(\alpha) = 0$, $t_1(\alpha) = 1$ *and*

$$t_{i+1}(\alpha) = \left\lceil \frac{k-i}{\beta} \right\rceil$$

where

$$\beta = \min \frac{e\alpha - v}{v},$$

the minimum over all positive integers v, e with $v \le t_i(\alpha)$ and $v - e\alpha < 0$. Their domain is the set of reals (both rational and irrational) $\alpha \in \left(\frac{1}{k-1}, 1\right)$.

Suppose α is irrational. The core argument of Section 6.1.2 defined the t_i for Duplicator's look-ahead strategy. These in turn depended on the Finite Closure Theorem 4.3.2. The look-ahead functions $t_i(\alpha)$ given above are precisely the t_i for the successful Duplicator look-ahead strategy. We may also express this in terms of the complete theory T_α, as given in Section 7.1. Let $\mathbf{x} = (x_1, \ldots, x_i)$ Let $P(\mathbf{x})$ be a predicate with bound variables $x_{i+1}, \ldots, x_k$. Set $t = t_i(\alpha)$. Then there is a finite list (possibly empty) $H_1, \ldots, H_s$ of possible t-closures such that

$$\forall_{\mathbf{x}} [P(\mathbf{x}) \leftrightarrow \bigvee_{i=1}^{s} \mathrm{cl}_t(\mathbf{x}) \cong H_i]$$

is a theorem of T_α.

9.2.3 Well Ordered Discontinuities

The functions $t_i(\alpha)$ have an intriguing behavior. Let us first illustrate this with an example. Let $k = 6$ and suppose $\alpha = \frac{1}{3} + \epsilon$ with ϵ small but positive. Consider $t_2(\alpha) = \lceil \frac{5}{\beta} \rceil$. The minimum for β is achieved when $v = 1, e = 3$ so that $\beta = 3\epsilon$. Then $t_2(\alpha) = \lceil \frac{5}{3\epsilon} \rceil$. As $\epsilon \to 0^+$, $t_2(\alpha) \to \infty$. The situation is not symmetric. When $\alpha = \frac{1}{3} - \epsilon$ we cannot take $v = 1, e = 3$ as then $v - e\alpha > 0$. In that case $v = 1, e = 4$ yields $\beta = \frac{1}{3} - \epsilon$ so that $t_2(\alpha) = \lceil \frac{5}{\beta} \rceil = 15$ for all small positive ϵ. This corresponds to our comments following the proof of the Finite Closure Theorem 4.3.2.

Claim 9.2.1 *For each $1 \le i \le k$ there is a set $\Omega_i \subset \left[\frac{1}{k-1}, 1\right]$ such that*

1. *$t_i(\alpha)$ is Left Continuous – For all $x \in \left(\frac{1}{k-1}, 1\right]$ there exists $\epsilon > 0$ so that $t_i(\alpha)$ is constant on $[x - \epsilon, x]$.*
2. *$t_i(\alpha)$ is continuous outside Ω_i – For all $x \in \left(\frac{1}{k-1}, 1\right)$ with $x \notin \Omega_i$ there exists $\epsilon > 0$ so that $t_i(\alpha)$ is constant on $[x - \epsilon, x + \epsilon]$.*
3. *The elements of Ω_i are all rational numbers.*
4. *Ω_i is well ordered under $>$.*
5. *Under the well ordering Ω_i has order type at most $c\omega^{i-1}$ for some c dependent on k, i.*
6. *$1, \frac{1}{k-1} \in \Omega_i$*

Further $\Omega_1 \subseteq \Omega_2 \subseteq \ldots \subseteq \Omega_k$.

Proof: We take $\Omega_1 = \left\{\frac{1}{k-1}, 1\right\}$. We use induction on i. For each $a \in \Omega_i$, $a \ne \frac{1}{k-1}$, let a^- (by the well ordering) denote the maximal element of Ω_i which is less than a. Set $I = (a^-, a]$. Then t_i has a constant value, say s, on I. On I we may write

$$t_{i+1}(\alpha) = \max \left\lceil \frac{(k-i)v}{e\alpha - v} \right\rceil,$$

the maximum over those integers v, e with $v \le s$ and $v - e\alpha < 0$. There are only finitely many potential pairs v, e. For each the condition $v - e\alpha < 0$ is equivalent to the lower bound $\alpha > \frac{v}{e}$. Split I by the finite set of all such $\frac{v}{e}$. This breaks I into a finite number of intervals $J = (c, d]$. By induction all endpoints c, d remain rational. Further, on each J the maximum defining $t_{i+1}(\alpha)$ is over a particular finite set of pairs v, e. For any particular v, e the function $\lceil (k-i)v/(e\alpha - v) \rceil$ will have discontinuities at the values

$$\alpha = \frac{v}{e} + \frac{v(k-i)}{ey}$$

where y is a positive integer. For v, e with $\frac{v}{e} < c$ we have only a finite number of these discontinuities in J. But in the critical case when $\frac{v}{e} = c$ we have a set of discontinuities of order type (under $>$) ω. We define Ω_{i+1} to be Ω_i

plus all endpoints c, d of the intervals J plus all discontinuities of t_{i+1} in the interior of each J. The endpoints c, d add only a finite set between any pair of consecutive points of Ω_i giving a set still of size $c'\omega^{i-1}$. The discontinuities of t_{i+1} now add a set of order type ω (or smaller) between any two consecutive values so that Ω_i has order type at most $c'\omega^i$.

Recall that the Ω_i, $1 \leq i \leq k$, did also depend on k. Now, to emphasize the dependence on k, we set $\Gamma_k = \Omega_k$.

Theorem 9.2.2. *Let A have quantifier depth k. All points α of discontinuity of f_A lie in Γ_k.*

We outline the argument. Suppose $I = [a, b)$ is an interval on which t_0 has constant value. Let $\alpha \in I$, irrational. We follow the argument in Section 7.1 that T_α is generated by the NonExistence and Generic Extension Schema. At each stage the notions of dense and sparse extensions for the needed extensions are uniform over $\alpha \in I$.

9.2.4 Underapproximation sequences

Definition 9.3 *Let $\alpha \in (0, 1)$, rational or irrational. We define the* underapproximation sequence *of α to be $\alpha_0, \alpha_1, \cdots$ where $\alpha_0 = 0$ and α_{i+1} is the maximal rational number $\frac{c}{d} \leq \alpha$ with*

$$d \leq \frac{10}{\alpha - \alpha_i}$$

When $\alpha_k = \alpha$ the sequence terminates and we call k the underapproximation length *of α. We set Δ_k equal those α with underapproximation length at most k.*

The value 10 in the above definition is simply a convenience and could be replaced by, say, 2 giving basically the same results. Clearly, the underapproximation sequence for an irrational α cannot terminate.

Claim 9.2.3 *All rational $\alpha \in (0, 1)$ have finite underapproximation length.*

Proof: Suppose

$$\frac{1}{t+1} < \alpha - \alpha_i \leq \frac{1}{t}$$

where α_i is as in Definition 9.3 above. Then $d = 10t$ is an allowable denominator for α_{i+1}. With this denominator we can get within $(10t)^{-1}$ of α so that

$$\alpha - \alpha_{i+1} \leq \frac{1}{10t}$$

Consider α_1: As $\alpha < 1$ and $\alpha_0 = 0$, $d = 10$ is a permissible denominator. Thus $\alpha_1 \geq \lfloor 10\alpha \rfloor / 10$ so $\alpha - \alpha_1 < \frac{1}{10}$. The values $\lfloor (\alpha - \alpha_i)^{-1} \rfloor$ increase at least tenfold for each i so that for some $i \leq \log_{10} d$ they reach d and the underapproximation sequence terminates with $\alpha_{i+1} = \alpha$.

Consider, for example, α slightly greater than $\frac{1}{3}$. These will have $\alpha_1 = \frac{1}{3}$. Suppose further $\alpha = \frac{1}{3} + \frac{1}{3s} = \frac{s+1}{3s}$. Then $\alpha_2 = \alpha_1$. Thus Δ_2 contains an infinite set. Further, consider $\alpha = \frac{1}{3} + \frac{1}{3s} + \frac{1}{3st}$. To avoid trivial cases think of s large and then t much larger. Then $\alpha_2 = \frac{1}{3} + \frac{1}{3s}$. As $(\alpha - \alpha_2)^{-1} = 3st$ we have $\alpha_3 = \alpha$. This places in Δ_3 a well ordered (under $>$) set of order type ω^2.

Claim 9.2.4 *For every k there exist l, $\epsilon > 0$ so that*

$$\Gamma_k \subseteq \Delta_l \cap [\epsilon, 1]$$

Further, for every l, $\epsilon > 0$ there exists k so that

$$\Delta_l \cap [\epsilon, 1] \subseteq \Gamma_k$$

For a given k, l and α we let $\alpha_i = v_i/e_i$ denote the underapproximation sequence to α (while defined) and $\alpha_i^* = v_i^*/e_i^*$ denote the maximal rational $\leq \alpha$ with numerator at most $t_i(\alpha)$. Call the α_i^* the look-ahead sequence of α. The idea of the proof is to show that the look-ahead and underapproximation sequences approach α at comparable speeds.

Let k be given. We set $\epsilon = \frac{1}{k-1}$. Suppose $\alpha = \frac{c}{d} \in \Gamma_k$. We wish to bound the underapproximation length of α. From the definition 9.2 of the look-ahead functions

$$t_{j+1}(\alpha) = \left\lceil \frac{k-j}{\beta} \right\rceil \leq \left\lceil \frac{k}{\beta} \right\rceil$$

where

$$\beta = \min \frac{\alpha - \frac{v}{e}}{\frac{v}{e}}$$

the minimum over all positive integers v, e with $v \leq t_j(\alpha)$ and $\frac{v}{e} < \alpha$. We first argue roughly that $\frac{v}{e}$ cannot be too small. Let s be such that $\frac{1}{s+1} \leq \alpha < \frac{1}{s}$. Then $v = 1, e = s$ satisfies the conditions and gives $\beta < 1$. As $\alpha \geq \frac{1}{k-1}$ any v, e with $\frac{v}{e} \leq \frac{1}{2(k-1)}$ would give a smaller value. Thus

$$\beta \geq 2(k-1) \min\left(\alpha - \frac{v}{e}\right)$$

and

$$t_{j+1}(\alpha) \leq \left\lceil 2k\,(k-1) \max\left(\alpha - \frac{v}{e}\right)^{-1} \right\rceil$$

Let us define the souped up look-ahead sequence α_i^{*+}. Given α_i^{*+} we set α_{i+1}^{*+} to be the maximum rational $\leq \alpha$ with numerator at most $2k\,(k-1)(\alpha - \alpha_i^{*+})^{-1}$. If $\alpha \in \Gamma_k$ then the souped up look-ahead sequence reaches α in at most $k+1$ steps. The souped up look-ahead sequence differs from the underapproximation sequence in that an additional factor of $k(k-1)$ is allowed in the numerator. But in the underapproximation sequence the numerator bound increases at least tenfold each time. Hence the underapproximation sequence will reach α in $O(k \ln k)$ steps.

Conversely, suppose l, ϵ are given. We choose k so that $\frac{1}{k-1} \leq \epsilon$ and $k \geq l+5$ (not best possible). Consider the first l steps of the look-ahead sequence. In each $k-i \geq 2$. Further in the minimum giving β all $v/e < \alpha \leq 1$ so that $\beta \geq \min(\alpha - \frac{v}{e})$. Thus the look-ahead function $t_{j+1}(\alpha) \geq 2(\alpha - \frac{v}{e})^{-1}$ where we can take $v/e = v_j/e_j$. Let $\alpha \in \Delta_l \cap [\epsilon, 1]$. The underapproximation sequence reaches α in at most l steps. The first l steps of the look-ahead sequence are at least as rapid and so they too reach α and $\alpha \in \Gamma_k$.

We may thus rewrite Theorem 9.2.2.

Theorem 9.2.5. *For any A the discontinuities of f_A are bounded away from zero and have bounded underapproximation length.*

The notion of bounded underapproximation length is an intriguing one, with interesting connections to the study of continued fractions.

9.2.5 Determination in PH

The set Γ_k is well ordered under $>$. For every $a \in \Gamma_k$ there is a unique next value, denote it by a^-. (When $a = \frac{1}{k-1}$, the last value of Γ_k under $>$, we set $a^- = 0$.) Every irrational $\alpha \in (0,1)$ belongs to a unique interval (a^-, a). Hence f_A is determined by giving its constant value on each such interval $I = (a^-, a)$.

Definition 9.4 *Let A be a first order sentence. We define the* rational limit function *of A, denoted by $g_A(\alpha)$ by*

$$g_A(\alpha) = \lim_{\beta \to \alpha^-} f_A(\beta)$$

Equivalently,

$$g_A(\alpha) = \begin{cases} 1 & \textit{if } A \in T_\alpha^- \\ 0 & \textit{if } A \notin T_\alpha^- \end{cases}$$

The domain of g_A is the set of rational $\alpha \in [0,1)$.

Here T_α^- is as defined in Section 8.4 and the equivalence of the above definitions is given by Theorem 8.4.1. The constant value of f_A on $I = (a^-, a)$ is then $g_A(a^-)$. Thus f_A is determined by the function g_A. Indeed, it is determined by g_A on $\Gamma_k \cup \{0\}$ but we shall not use this fact. Our next result limits the complexity of the function g_A.

Theorem 9.2.6. *For any A the set of rational $\alpha \in (0,1)$ such that $A \in T_\alpha^-$ is in Polynomial Time Hierarchy complexity class PH. Here we write a rational $\alpha = \frac{c}{d}$ in unary so that α has length $\Theta(d)$.*

We determine if $A \in T_\alpha^-$ by employing the Truth Game of Section 7.1.3. We think of A, and hence the quantifier depth k, as fixed and look at the complexity in terms of d, where $\alpha = \frac{c}{d}$. We examine the $t_0 = 0, t_1 = 1, \ldots, t_k$ for the look-ahead strategy. Then, as given in Section 8.4.1

$$t_{i+1}(\alpha) = \left\lceil \frac{k-i}{\beta} \right\rceil$$

where

$$\beta = \min \frac{e\alpha - v}{v},$$

the minimum over all integers v, e with $v \le t_i(\alpha)$ and $v - e\alpha < 0$.

Since $v - e\alpha$ must be strictly negative and v, e are integral we must have

$$v - e\alpha \le -\frac{1}{d}$$

so that

$$\beta \ge \frac{1}{dt_i(\alpha)}$$

and therefore

$$t_{i+1}(\alpha) \le (k-1)dt_i(\alpha) + 1$$

With k fixed we therefore can write

$$t_i(\alpha) = O(d^{i-1})$$

In particular, all $t_i(\alpha)$ are of polynomial size in d.

Now the positions H of the Truth Game (the possible t_i-closures of $k-i$ vertices) can can be described by a polynomial length (in d) string of zeroes and ones, giving the incidence matrix of H and the marked vertices. Let H, H' be positions on the i and $i+1$ levels respectively. What is the complexity of determining if (H, H') is an acceptible move – that is, following Claim 7.1.3, determining if

$$\forall_{\mathbf{x}}\, \mathrm{cl}_u(\mathbf{x}) \cong H \rightarrow \exists_x\, \mathrm{cl}_t(\mathbf{x}, x) \cong H_1$$

is a theorem in T_α^-. We claim this complexity is in PH, in terms of the sizes of H, H' and u, t.

Suppose this is a theorem in T_α^-. Verifier gives an $H^* = \mathrm{pict}(H, H_1)$ (a polynomial string), with $\mathbf{x}$,H, H_1, x all designated.

First suppose $x \in H$, the Inside case, so that $H = H^*$. She also gives a polynomial string of rigid extensions from $\mathbf{x}, x$ to reach all of H_1. Now Spoiler checks each such extension (R, R'). From Property 4.1.8 a nonrigid extension would have a safe, hence sparse, nailextension. Inversely, rigid extensions have all nailextensions rigid, hence dense. If (R, R') is not rigid Spoiler gives an S, $R \subseteq S \subset R'$ with the nailextension (S, R') sparse. He gives S with a polynomial string and then checking for sparseness can be done in polynomial time. It may also be that H_1 is not all of $\mathrm{cl}_t(\mathbf{x}, x)$. In that case Spoiler gives a polynomial string giving a rigid extension (H_1, H_2) with $H_2 \subset H$ with less than t nonroots. To be sure, (H_1, H_2) might not actually be rigid but if it isn't Verifier supplies a nail extension S so that (S, H_2) is sparse.

Now suppose $x \notin H$, the Outside case. Verifier first demonstrates that in H^* the t-closure of $\mathbf{x}, x$ is H_1. As in the previous paragraph she gives a polynomial string of rigid extensions from $\mathbf{x}, x$ to reach all of H_1. Spoiler's response and Verifier's counterresponse are as in the above paragraph. Let NEW be the vertices of H^* not in H. Verifier claims that NEW is a safe extension over H. If not, by Property 4.1.8 there is a rigid, hence dense, subextension and if so all subextensions are sparse. If NEW is not safe over H then Spoiler gives the dense subextension S with a polynomial string and checking for denseness is done in polynomial time.

In all cases the determination of whether (H, H') is an acceptible move is done with a bounded number of alternations and so the query is in the complexity class PH.

But now we have Theorem 9.2.6. The Truth Game requires a finite number of moves each of which is a polynomial length string. The validity of the move can be challenged but this takes a bounded number of moves which are polynomial strings, reducing all to polynomial computation. This places the query "$A \in T_\alpha^-$?" in the hierarchy PH.

10. Strings

Here we look at strings over a finite alphabet, such as *abbacab* or *bbaacccb*. Our sentences can use words like before, after, next, first, last and would include such possibilities as "Every a is immediately followed by a b," or "Between every two bs lies a c."

10.1 Models and Language

Let Σ be a finite alphabet, $\Sigma = \{0, 1\}$ and $\Sigma = \{a, b, c\}$ being natural examples. Let Σ^* denote, as customary, the set of finite strings from Σ, such as 0111101 or *abbacca* in the above examples. We include the empty string, denoted O, in Σ^* – it plays a role analogous to the null set. The basic operation in Σ^* is concatenation, denoted by $w + w'$. Thus $abba + bca = abbabca$. Observe $+$ is associative (but not Abelian!) and O is an identity so that Σ^* forms a semigroup. We naturally define, for s a positive integer, sw to be w concatenated with itself s times, so that $4(ab) = abababab$.

Definition 10.1 *The* linear language *for Σ consists of equality, unary predicates U_α for each $\alpha \in \Sigma$ and the binary relation less than, written $x < y$.*

The term linear is used to distinguish from the circular language of Definition 10.9 and refers to $<$, which we interpret as a linear order. A finite model M of the language consists of an ordered set of some n elements, which we shall generally denote $1, \ldots, n$, with the unary predicates U_α defined on it so that for every x precisely one of the $U_\alpha(x)$ holds. The finite models M are in a natural bijective correspondence with the $w \in \Sigma^*$. Concatenation of strings goes over to concatenation $M + M'$ of models.

Now we can say for $\Sigma = \{a, b, c\}$:

- Every a is immediately followed by a b

$$\forall_x \Big[U_a(x) \rightarrow \exists_y \big[U_b(y) \wedge \neg\exists_z (x < z \wedge z < y)\big]\Big]$$

- Between every two bs lies a c

$$\forall_{x,y} \Big[\big(U_b(x) \wedge U_b(y) \wedge x < y\big) \rightarrow \exists_z \big(x < z \wedge z < y \wedge U_c(z)\big)\Big]$$

- The first letter is a c

$$\exists_x \left[U_c(x) \wedge \neg\exists_y (y < x)\right]$$

and much else. Note that the use of less than makes this a much more powerful language than the simple unary predicate of Section 0.2.

10.2 Ehrenfeucht Redux

10.2.1 The Rules

Analogously to Chapter 2 (and given in full generality in Section 2.6.1) we define the k-move Ehrenfeucht game $\textsc{Ehr}(M_1, M_2; k)$ between Spoiler and Duplicator. Each round has two parts, Spoiler's move followed by Duplicator's move. On the i-th move Spoiler selects a vertex in either model (his choice) and marks it i. Then Duplicator must select a vertex in the other model and also mark it i. A vertex may receive more than one mark.

Who wins? At the end of the game let $x_1, \ldots, x_k$ be the vertices of M_1 marked $1, \ldots, k$ respectively [regardless of who put the mark there] and let $y_1, \ldots, y_k$ be the corresponding vertices of M_2. For Duplicator to win she must assure that, for all $1 \le i < j \le k$, $x_i < x_j$ if and only if $y_i < y_j$ and $x_i = x_j$ if and only if $y_i = y_j$. She must all assure the for each $1 \le i \le k$ and each $\alpha \in \Sigma$ that $U_\alpha(x_i)$ if and only if $U_\alpha(y_i)$ If Duplicator doesn't win then Spoiler wins.

It may be helpful to think of the game on the strings. Spoiler picks a position on one string. Duplicator must pick the same letter on the other string and keep intact the order of the choices. For example, consider the two move game on *abcba* and *acbca*. Spoiler first selects (among many winning strategies) the first b on the first string: ab^1cba. Duplicator must reply by selecting a b: acb^1ca. Spoiler then selects the first c on the second string ac^2b^1ca. While Duplicator has a c to select: ab^1c^2ba it is not on the same side of the first choice and so she loses.

We write $M_1 \equiv_k M_2$ if Duplicator wins $\textsc{Ehr}(M_1, M_2; k)$. We repeat Theorem 2.6.1 in this important special case:

Theorem 10.2.1. *For all $k \ge 1$:*

1. *$M_1 \equiv_k M_2$ if and only if M_1, M_2 have the same truth value for all sentences in the linear language quantifier depth k.*
2. *To each k-Ehrenfeucht value there is a sentence A in the linear language of quantifier depth k such that the models M with that value are precisely those M for which A is true.*

10.2.2 The Semigroup

Theorem 10.2.2. *If $M_1 \equiv_k M_1'$ and $M_2 \equiv_k M_2'$ then $M_1+M_2 \equiv_k M_1'+M_2'$.*

Proof: Consider the game on the two models $M_1 + M_2$ and $M_1' + M_2'$. When Spoiler selects an $x \in M_1$ Duplicator pretends that she is playing the k-move game on M_1 and M_1' and makes the corresponding reply on M_1'. When Spoiler selects an $x \in M_2$ Duplicator pretends that she is playing the k-move game on M_2 and M_2' and makes the corresponding reply on M_2'. Plays on M_1' and M_2' are treated similarly. Duplicator is then always preserving the unary predicates (the letter) and the order inside the submodels. But if $x \in M_1$ and $y \in M_2$ have been selected then their analogous vertices also have $x' \in M_1'$ and $y' \in M_2'$. Since *every* element $x \in M_1$ is less than every $y \in M_2$ (and also for M_1', M_2') Duplicator has preserved order in these cases as well.

Definition 10.2 *The* Ehrenfeucht Semigroup $\text{EHR}[k]$ *is the set of all k-equivalence classes of finite models M under the induced operation of concatenation.*

That is, if m_1, m_2 are equivalence classes and M_1, M_2 representatives of that class, then $m_1 + m_2$ is the equivalence class containing $M_1 + M_2$. The associativity of concatenation on models gives the associativity of concatenation on $\text{EHR}[k]$. Let $\alpha \in \Sigma$. We shall use α to denote four similar objects. First, the element of Σ. Second, the one letter string α which is an element of Σ^*. But $w \in \Sigma^*$ have a natural correspondence with finite models so this gives a third meaning: the model α is on the single element 1 with $U_\alpha(1)$. Finally, we let α represent the element of $\text{EHR}[k]$ which is the k-equivalence class containing the model α. For $k \geq 2$ this class is particularly simple. Consider the two move Ehrenfeucht game with models α and M. If in M we have $U_\beta(x)$ for any x and any $\beta \neq \alpha$ then Spoiler wins in one move by selecting that x. If M has two (or more) distinct elements x, y then Spoiler wins by selecting $x, y \in M$. Thus for $k \geq 2$ the equivalence class $\alpha \in \text{EHR}[k]$ consists simply of the model α itself. Similarly, we have defined O to be the empty string. We also let O denote the model on no elements. We also let $O \in \text{EHR}[k]$ be the k-equivalence class containing the model O. For $k \geq 1$ this class consists simply of the model O itself. Observe that $O \in \text{EHR}[k]$ serves as an identity, $m + O = m$ as concatenating with the empty string does not change a model.

Analogously to Theorem 2.2.1 we have

Theorem 10.2.3. $\text{EHR}[k]$ *is a finite set.*

10.2.3 Long Strings

Theorem 10.2.4. *For $s, t \geq 2^k - 1$ and any $w \in \Sigma^*$*

$$sw \equiv_k tw$$

Proof: We use Theorem 2.6.3. Say Spoiler plays in the a-th copy of w in sw. Duplicator imagines that he has played $a \in \overline{s}$ in the total order game on models $\overline{s}, \overline{t}$. In the total order game she has a response $b \in \overline{t}$. She then goes to the b-th copy of w in tw. There she plays the exact same place in w that Spoiler did. Spoiler might play several times in the a-th copy of w. Each time Duplicator will respond in the same b-th copy of w making the identical move.

Property 10.2.5 *In the Ehrenfeucht Semigroup* EHR$[k]$

$$sw = tw$$

for all w and all $s, t \geq 2^k - 1$.

10.3 Persistent and Transient

The results of this section hold in any finite semigroup M with identity and with Property 10.2.5. Their interpretation for the Ehrenfeucht Semigroup EHR$[k]$ is given at the end of the section. The use of letters p, s below is meant to suggest prefix and suffix respectively.

Theorem 10.3.1. *For $x \in M$ the following three properties are equivalent:*

1. $\forall_y \exists_z (x + y + z = x)$
2. $\forall_y \exists_z (z + y + x = x)$
3. $\exists_p \exists_s \forall_y (p + y + s = x)$

Definition 10.3 *We call $x \in M$* persistent *if it has the properties of Theorem 10.3.1. We call $x \in M$* transient *if it is not persistent.*

Proof of Theorem 10.3.1:
$3 \Rightarrow 1$: Take $z = s$, regardless of y. Then

$$x + y + z = (p + y + s) + y + s = p + (y + s + y) + s = x$$

$1 \Rightarrow 3$: Set $R_x = \{x + v : v \in M\}$ so that $R_x + u = \{x + v + u : v \in M\}$. We first claim there exists $u \in M$ with $|R_x + u| = 1$, i.e., all $x + y + u$ the same. Otherwise take $u \in M$ with $|R_x + u|$ minimal and say $v, w \in R_x + u$. As $R_x + u \subseteq R_x$ we write $v = x + u_1, w = x + u_2$. From 1, with $y = u_1$, we have $x = v + u_3$ and thus $w = (v + u_3) + u_2 = v + u_4$ with $u_4 = u_3 + u_2$. Now we set $S = 2^k - 1$ (the actual value doesn't matter) and apply Property 10.2.5.

$$w + Su_4 = v + (S+1)u_4 = v + Su_4$$

Adding Su_4 to $R+u$ sends v, w to the same element so $|R+u+Su_4| < |R+u|$, contradicting the minimality. Now say $R_x + u = \{u_5\}$. Again by 1 there exists u_6 with $u_5 + u_6 = x$. Then $R_x + (u + u_6) = \{x\}$ so that 3 holds with $p = x, s = u + u_6$.

By reversing addition (noting that 3 is selfdual while the dual of 1 is 2) these arguments give that 3 and 2 are equivalent.

We can define a directed graph G^r on M by directing edges from x to $x + y$ for every $x, y \in M$. From 1 the persistent elements of M are precisely the elements of the strongly connected components of G^r.

Property 10.3.2 *If z is persistent and $c \in M$ then*

$$z + c + z = z$$

Proof: Let p, s, as given by 3, have the property that $p + y + s = z$ for all y. Taking $y = O$, $p + s = z$. Then $z + c + z = p + (s + c + p) + s = z$.

Property 10.3.3 *If x is persistent then $w_1 + x + w_2$ is persistent for all $w_1, w_2 \in M$.*

Proof: Let x be persistent and set $v = w_1 + x + w_2$. For any $y \in M$ set $z = v$ so that

$$v + y + z = w_1 + \left[x + (w_2 + y + w_1) + x\right] + w_2 = w_1 + x + w_2 = v$$

and hence v is persistent.

Definition 10.4 *We define the relation $x \equiv_R u$ by $\exists_v(x + v = u)$. We define the relation $x \equiv_L u$ by $\exists_v(v + x = u)$.*

Property 10.3.4 *$\equiv_R$ and $\equiv_L$ are equivalence relations on the set of persistent elements of M.*

Proof: Immediate from properties 1, 2 respectively of Theorem 10.3.1.

Definition 10.5

$$R_x = \{x + v : v \in M\}$$
$$L_x = \{v + x : v \in M\}$$

For x persistent R_x is the strongly connected component of G^r containing x.

Property 10.3.5 *For x persistent R_x is the equivalence class containing x under $\equiv_R$ and L_x is the equivalence class containing x under $\equiv_L$.*

Proof: Immediate.

Property 10.3.6 *Let x, y be persistent. Then*

$$R_x \cap L_y = \{x + y\}$$

Proof: Clearly $x + y \in R_x \cap L_y$. Let $z \in R_x \cap L_y$. Then there exist a, b with $x = z + a$ and $y = b + z$ so that $x + y = z + (a + b) + z$. By Property 10.3.3 z is persistent and by Property 10.3.2 $z + (a + b) + z = z$, so that $x + y = z$.

Property 10.3.7 *Let w be persistent. Then*

$$p + w + m + w + s = p + w + m' + w + s$$

for all p, s, m, m'.

Proof: By Property 10.3.3 both $x = p + w$ and $y = w + s$ are persistent. Thus $p + w + m + w + s = x + m + y = x + y$ by Property 10.3.6.

One final property shows that these ideas are not pointless.

Property 10.3.8 *There exists a persistent $x \in M$.*

Proof: Take x with $x + M = \{x + m : m \in M\}$ of minimal size. For any $y \in M$ $x + y + M \subseteq x + M$ so that $x + y + M = x + M$. As $x = x + O \in x + M$, $x \in x + y + M$ so that there exists z with $x + y + z = x$ and x is persistent.

10.4 Persistent Strings

We fix a finite alphabet Σ and the parameter k of the previous section.

Definition 10.6 *We call $w \in \Sigma^*$* k-persistent *if its Ehrenfeucht value m is persistent as given by Definition 10.3. Otherwise, we call w* k-transient. *We use the terms persistent and transient when k is understood.*

Our object in this section is to get a reasonable picture of what a persistent strings is. Our first result is a sufficiency condition. Following standard usage we say that w is a *substring* of τ if it lies inside τ as consecutive letters, with no gaps. Equivalently, $\tau = p + w + s$ for some strings p, s.

Theorem 10.4.1. *There is a string w so that any string τ containing w as a substring is persistent.*

Proof: Use Property 10.3.8 to pick any w whose Ehrenfeucht value m is persistent. Any $\tau = p + w + s$ has Ehrenfeucht value $m^- + m + m^+$, where m^-, m^+ are the Ehrenfeucht values of the strings p, q respectively. But $m^- + m + m^+$ is persistent by Property 10.3.3.

Definition 10.7 *A first order sentence B (in the first order language for Σ) is called* central *if (i) some $w \in \Sigma^*$ satisfies B and (ii) if w satisfies B and τ contains w as a substring then τ satisfies B.*

Theorem 10.4.2. *w is k-persistent if and only if it satisfies all central sentences B of quantifier depth at most k.*

Proof: Assume w persistent and B central. Let v satisfy B. Then $w + v + w$ satisfies B. But $w + v + w$ has the same Ehrenfeucht value as w so they have the same truth value on B and so w satisfies B.

Now for the converse. For each Ehrenfeucht value α let A_α be that first order property that holds if and only if the string has that value. The existence of A_α is given by Theorem 10.2.1. Let B be the disjunction of the A_α over the persistent states α. We claim B is a central property. If w satisfies B then w satisfies some particular A_α and so w is persistent. From Property 10.3.3 any τ containing w as a substring is persistent. Further, from Property 10.3.8 there are w satisfying B, hence B is indeed a central property. If w satisfies all central properties of quantifier depth at most k then it must satisfy this particular property B. But we've just shown that if w satisfies B then w is indeed persistent.

Roughly, the central sentences are existential statements that do not depend on the "edges" of a string. Let B be the sentence that the string begins with a. This sentence is not central: if you take a string starting with a and add a b on the left then it no longer has this property. Similar noncentral properties would be that the first non-a is a c or that the first time either aba or aca occurs as a substring it is aba. All of these can have their truth value changed by changing the edges of the string. A typical central statement is that $accab$ appears as a substring. Once it does no additions to the string on the edges can make it false. A more complicated statement is that there exist two a's with precisely one b between them. Again, once this appears it cannot be destroyed by adding to the string at the edges. Thus a persistent string w (when $k \geq 4$, the quantifier depth of this sentence) *must* have two a's with precisely one b between them.

10.5 Random Strings

Let us fix a nontrivial distribution on Σ. That is, for each $\alpha \in \Sigma$ we are given p_α positive with $\sum p_\alpha = 1$. Let Σ^n denote the strings from Σ of length n. Then we may naturally give Σ^n the product distribution, each letter being independently selected from the above distribution. For a property A we can set $\mu_n(A)$ equal the probability that the randomly selected $w \in \Sigma^n$ has property A. The following early result was inspirational to many.

Theorem 10.5.1 (Ehrenfeucht). *For any first order property A*

$$\lim_{n\to\infty} \mu_n(A) \text{ exists}$$

Proof: Let A have quantifier depth k. We create a Markov Chain! The states are the Ehrenfeucht values $\text{EHR}[k]$. The initial state is O. Recall that we consider $\alpha \in \Sigma$ as being in $\text{EHR}[k]$, corresponding to the one letter string α. Now suppose $P, Q \in \text{EHR}[k]$ and $P+\alpha = Q$. We set the one step transition probability from P to Q as p_α. (With $k \geq 2$ we never have $P + \alpha = P + \beta$ as they have different last letters.) Let $p^{(n)}(P, Q)$ denote, as usual in Markov

Chain theory, the n-step transition probability from P to Q. For each $Q \in$ EHR$[k]$ either all strings in class Q satisfy A or no strings in class Q satisfy A. Let $S \subset$ EHR$[k]$ be the set of those Q for which all strings in class Q satisfy A. Then $\mu_n(A)$ is the probability that after n steps the Markov Chain is in set S. That is,

$$\mu_n(A) = \sum_{Q \in S} p^{(n)}(O, Q)$$

It suffices therefore to show for any given $Q \in$ EHR$[k]$ that $\lim_{n\to\infty} p^{(n)}(O, Q)$ exists.

Now we employ a basic (though nontrivial) theorem on Markov Chains. The only way this limit can not exist is if the Markov Chain exhibits a periodic behavior – in particular if there exists $d > 1$ and disjoint sets $E_0, E_1, \ldots, E_{d-1}$ of states, with $Q \in E_0$, such that from any state in E_i in one step you must go to a state in E_{i+1}, and from E_{d-1} you must go to E_0. For any $\alpha \in \Sigma$ we would then have $Q + i\alpha \in E_0$ if and only if $d \mid i$. Now we make critical use of Property 10.2.5 and give a contradiction, as for i sufficiently large $Q + i\alpha = Q + (i+1)\alpha$. Hence the limit does indeed exist.

The terms persistent and transient were in fact taken from Markov Chain theory. An Ehrenfeucht value α is persistent precisely if it is persistent as a state in the Markov Chain defined above. Let TR be the property that the Ehrenfeucht value of a string w is transient (relative to a fixed k).

Theorem 10.5.2. *There exist constants $K > 0$ and $\epsilon > 0$ so that*

$$\mu_n(TR) < K(1-\epsilon)^n$$

Proof: We can deduce this from general results about Markov Chains. Alternatively, we know Property 10.3.8 that there is a persistent string x. Suppose it has length l and the probability of a random string of length l being x is ϵ'. Parse a random string w of length n into $n' = \lfloor n/l \rfloor$ segments of length l. If any of these segments is x then, by Property 10.3.3, w is persistent. The probability of that not occurring is $(1-\epsilon')^{n'}$. As $n' \geq \frac{n}{l} - 1$ this is at most $K(1-\epsilon)^n$ where $K = (1-\epsilon')^{-1}$ and $1 - \epsilon = (1-\epsilon')^{1/l}$.

10.6 Circular Strings

Theorem 10.5.1 cannot be a Zero-One Law. The simple property that the first letter is an a has limiting probability p_a which is neither zero nor one. However, the results of Section 10.3 have indicated that the difficulties lie only in the edge effects – that we are able to make statements about the letters near the left and right edges. Now we move to a new model where those effects are eliminated.

Definition 10.8 *A circular string of length n over alphabet Σ is a function $f\colon Z_n \to \Sigma$.*

Definition 10.9 *The* circular language *over Σ consists of unary predicates U_α for each $\alpha \in \Sigma$, equality, and a ternary $C(x,y,z)$. On a circular string $f\colon Z_n \to \Sigma$ we interpret $U_\alpha(x)$ by $f(x) = \alpha$ and $C(x,y,z)$ by $x < y < z$ or $y < z < x$ or $z < x < y$.*

We have removed less than and replaced it with C which means "in a clockwise direction." We may still speak of y being the next element after x: $\neg\exists_z C(x,z,y)$. But we no longer can think of y as coming before or after x. We have eliminated the notions of first or last, we no longer have the property that the first letter is a.

Given a distribution p_α we naturally define a random circular string of length n. For any A in the circular language we set $\mu_n(A)$ equal the probability that the random circular string of length n has property A. We emphasize that the p_α are not (unlike Section 10.7 below) dependent on n so that our situation is analogous to the Fagin-GKLT Theorem 0.1.2 for arbitrary, but fixed, edge probability p.

Theorem 10.6.1. *In the circular language for any A*

$$\lim_{n\to\infty} \mu_n(A) = 0 \text{ or } 1$$

For a circular string $f\colon Z_n \to \Sigma$ and any $u \in Z_n$ we may straighten f out by listing its letters in ascending order (with 0 following $n-1$) beginning at u. Formally we define $f_u\colon \{1,\ldots,n\} \to \Sigma$, setting $f_u(x) = f(u-1+x)$ where the addition is done in Z_n. Now we give a definition analogous to Definition 2.3.

Definition 10.10 *A circular string $f\colon Z_n \to \Sigma$ has the k-Alice's Emporium property if*

1. *For every position $u \in Z_n$ there exist strings p, m, s so that $f_u = p+m+s$ and p, s are persistent.*
2. *For every pair of persistent strings p, s there exists a position $u \in Z_n$ and strings p', m, s' so that $f_u = p'+m+s'$ and $p' \equiv_R p$ and $s' \equiv_L s$ as given by Definition 10.3.4.*

Claim 10.6.2 *If circular strings $f\colon Z_n \to \Sigma$ and $g\colon Z_m \to \Sigma$ both have the k-Alice's Emporium property then they have the same truth values on all properties in the circular language with quantifier depth at most k.*

Proof: We play the Ehrenfeucht game. Spoiler first selects $x = x_1 \in Z_n$ (say). Duplicator expresses $f_x = p + m + s$ with p, s persistent. She selects $y = y_1 \in Z_m$ so that $g_y = p' + m' + s'$ with $p' \equiv_R p$ and $s' \equiv_L s$. From Property 10.3.6 f_x, g_y have the same Ehrenfeucht value. In particular (for $k \geq 2$) they have the same first letter so Duplicator has selected the same letter of Σ as Spoiler. Henceforth Spoiler plays as if it were the k-move (giving up a move) Ehrenfeucht game on f_x, g_y. Because they have the same

Ehrenfeucht value she wins. But has she won the circular game? Consider any triple x_i, x_j, x_l that have been chosen on Z_n and their corresponding y_i, y_j, y_l. Suppose $x_i < x_j < x_l$ in the straightened out ordering x to $n-1$ to 0 to $x-1$. Then as Spoiler preserved order we have $y_i < y_j < y_k$. Then $C(x_i, x_j, x_k)$ and $C(y_i, y_j, y_k)$. That is, Spoiler has actually also preserved the clockwise ternary relation and so she has won.

Claim 10.6.3 *There is a string* $U \in \Sigma^*$ *(dependent on* k*) such that any circular string containing* U *as a substring has the* k*-Alice's Emporium property.*

Proof: For each pair (P_i, S_i), $1 \leq i \leq R$ consisting of an $\equiv_R$ class followed by an $\equiv_L$ class let p_i, s_i be strings in those classes. Set

$$U = s_1 + p_1 + \ldots + s_R + p_R$$

Consider any circular string with U as a substring. Letting u be the first position of p_i we can write $f_u = p_i + m + s_i$ with $m = s_{i+1} + \ldots + p_{i-1}$. (The addends of m are written in circular order, with s_1 following p_R.) Now consider any position u. The straightening f_u cannot split both $s_1 + p_1$ and $s_2 + p_2$. We may write $f_u = x + s_i + p_i + y$ for some $i \in \{1, 2\}$. But $x + s_i$ and $p_i + y$ are persistent by Property 10.3.3 giving a parsing $f_u = p + m + s$ with $m = O$ and $p = x + s_i$, $s = p_i + y$ persistent.

Proof of Theorem 10.6.1: It suffices from Claim 10.6.2 to show that almost all w have the k-Alice's Emporium property and then from Claim 10.6.3 that almost all w have U as a substring. Indeed, the probability that a random w does not have a fixed U as a substring is exponentially small, as given by the proof of Theorem 10.5.2.

10.7 Sparse Unary Predicate

Let us set $\Sigma = \{0, 1\}$ and consider both the linear and circular language. As $U_0(x) \leftrightarrow \neg U_1(x)$ we may reduce to a single unary predicate (which we think of as a 1) which we'll write as U. We'll consider the random model of size n where $U(x)$ has probability p and call that $U(n,p)$. We may think of $U(n,p)$ as a random bit string, or a random circular bit string. Now, as in Section 0.2, we do not restrict ourselves to p constant but rather consider $p = p(n)$ as a function of n. We say a sentence A holds almost surely or almost never if the limit probability of it being satisfied is one or zero respectively.

What are the threshold functions in the evolution of the random bit string $U(n,p)$? At $p \sim n^{-1}$ a 1 appears. At $p \sim n^{-1/2}$ the substring 11 appears. Suppose $p \sim cn^{-1/k}$. The expected number of substrings of k consecutive ones is then asymptotically c^k and it can be shown that the probability that there is no such substring is asymptotically $\exp[-c^k]$. For our purposes here the key point is that this limit is neither zero nor one. For general $p = p(n)$ this limit probability is one if and only if $p(n) \ll n^{-1/k}$ and zero if and only if $p(n) \gg n^{-1/k}$.

Definition 10.11 *We say* $p(n)$ satisfies the Zero-One Law *for the circular language if, relative to it, every circular language sentence A holds either almost always or almost never. We say* $p(n)$ satisfies the Zero-One Law *for the linear language if, relative to it, every linear language sentence A holds either almost always or almost never.*

Edge effects make the Zero-Law Law for the linear language fail except in relatively simple cases. Consider the linear sentence A that there is an appearance of the substring 11 that occurs before all appearances of the substring 101. Suppose $n^{-1/2} \ll p(n) = o(1)$. One can show that almost surely both 11 and 101 are substrings and that each is equally likely to be the first, so that $\Pr[A] \to \frac{1}{2}$. This problem does not occur in the circular language.

If $p(n)$ satisfies the Zero-One Law for the circular language there must be limiting probability zero or one for the sentence "there exist k consecutive ones" and so either $p(n) \ll n^{-1/k}$ or $p(n) \gg n^{-1/k}$. Switching zero and one, either $1-p(n) \ll n^{-1/k}$ or $1-p(n) \gg n^{-1/k}$. These considerations force $p(n)$ to be in one of the five categories in our next result.

Theorem 10.7.1 (Shelah, Spencer[16]). $p(n)$ *satisfies the Zero-One Law for the circular language if and only if one of the following holds:*

1. $p(n) \ll n^{-1}$
2. *For some integer* r, $n^{-1/r} \ll p(n) \ll n^{-1/(r+1)}$
3. *For all* $\epsilon > 0$, $n^{-\epsilon} \ll p(n)$ *and* $n^{-\epsilon} \ll 1 - p(n)$
4. *For some integer* r, $n^{-1/r} \ll 1 - p(n) \ll n^{-1/(r+1)}$
5. $1 - p(n) \ll n^{-1}$

Why are these conditions sufficient? When $p(n)$ satisfies the third condition we can use Claim 10.6.3 as any fixed substring U will almost surely appear. Conditions four and five are equivalent to conditions two and one, replacing $U(x)$ by $\neg U(x)$. When $p(n)$ satisfies the first condition the situation is that of Section 3.1, there are no ones. We can therefore apply Theorem 10.2.4. The difficulty in Theorem 10.7.1 is then to show for each fixed r that the Zero-One Law is satisfied for $n^{-1/r} \ll p(n) \ll n^{-1/(r+1)}$. We do not prove Theorem 10.7.1 in this work but instead indicate the arguments for $r = 1$ and $r = 2$.

Suppose $n^{-1} \ll p(n) \ll n^{-1/2}$. We employ the analogue of the Bridge Theorem 2.5.1. Let k be fixed. Let U_n, U_m be independent random circular models of size n, m. It suffices to show that Duplicator almost always wins $\text{EHR}[U_n, U_m; k]$. For any fixed T the random U_n, U_m almost surely have more than T ones and no pairs of ones within T of each other. From Theorem 10.2.4, with $w = 0$, we can select T so that Duplicator wins the linear game $s0$ versus $t0$ for $s, t > T$. By a simple extension of Theorem 10.2.4 we can select T so that Duplicator wins the circular game $s1$ (i.e., $f\colon Z_s \to \{0,1\}$ with all $f(x) = 1$) versus $t1$ for $s, t > T$. When Spoiler selects a 0 we imagine that he has also selected the first 1 preceding it, in a counterclockwise direction. We don't count this extra play so this cannot help Duplicator. Duplicator responds first to the 1. She ignores all the 0s and plays on two circular boards composed only of 1s. With only k moves in the game she can always do this. She responds to the 0 by playing in the interval $s0$ following the 1. She might have to make many responses in the same interval, but at most k, and so she can always do this as well. One can check that with this strategy the clockwise relation is always preserved, as are the choices of 0 and 1.

Now suppose $n^{-1/2} \ll p(n) \ll n^{-1/3}$. Again fix the number of moves k of the Ehrenfeucht game over random U_n, U_m. Select T with the properties above. Call a substring $1 + i0 + 1$ with $0 \le i \le T$ a spike of type i. Spikes have appeared. Call the interval between two consecutive spikes a gap. One can show that all gaps have nice properties: They have at least T ones and between any two ones are at least T zeroes, including at the start and end of the string. That is, the gap behaves like the linear string with $n^{-1} \ll p(n) \ll n^{-1/2}$. We may list the types of the spikes in circular order, this gives us a circular string over $\Sigma = \{0, 1, \ldots, T\}$. The random U_n, U_m then give circular strings, call them $s(U_n), s(U_m)$. Let U^* be the universal string over Σ (for the given number of moves k) given by Claim 10.6.3. One can show that almost surely not only have spikes appeared but any particular finite string of spike types has appeared, so that $s(U_n), s(U_m)$ almost surely have U^* as a substring and hence are k-equivalent. Now Spoiler makes a move. If it is in a gap we give Spoiler a free move: the last 1 of the last spike preceding his move. Spoiler has made a spike move and possibly a gap move. Duplicator responds first to the spike move, since she can win the game of $s(U_n), s(U_m)$ she finds an equivalent spike in the other model and as the type of the spike tells exactly what it is she knows exactly where in the spike to play. To the gap move she must play in the particular gap following her spike reply. She will have to play at most k times in any particular gap but this she can do as all gaps are k-equivalent.

The Zero-One Law of Theorem 10.7.1 leads to a complete theory and some interesting countable models. These have been explored in some detail by Katherine St. John and the author [20]. Here we give only a rough picture. We look at the intervals where the probability is $p(n) \ll n^{-1}$ (almost surely no ones occur), $n^{-1} \ll p(n) \ll n^{-1/2}$ (only isolated ones occur), and $n^{-1/2} \ll p(n) \ll n^{-1/3}$ (adjacent ones occur). Moving from one interval to the next, the complexity of the model of the almost sure theory increases. Construction of the model for each theory leans heavily on the models of the previous theory. The models are over Z^k with the usual order: $\mathbf{x} = (x_1, \ldots, x_k) < \mathbf{y} = (y_1, \ldots, y_k)$ if, for some $1 \le i < k$, $x_j = y_j$ for $j < i$ and $x_i < y_i$. From this order the ternary $C(x, y, z)$ is defined by "$x < y < z$ or $y < z < x$ or $z < x < y$." Note that while the order is used as an auxiliary relation to define the model it does not appear in the model itself.

Suppose first $p(n) \ll n^{-1}$ so that $\forall_x \neg U(x)$ is in the theory. The unary predicate is therefore out of the picture. One needs only a set with a ternary clockwise C. The set must be infinite and every element needs a successor and predecessor. The simplest model is given by the integers Z. We think of Z with the usual $<$ and no position one as an ordinary line. The ordinary line (with the induced ternary C) provides the simplest model. We further think of Z with the usual $<$ and a single one in position zero as a distinguished line.

Now say $n^{-1} \ll p(n) \ll n^{-1/2}$. In a countable model one needs an infinite number of ones but between any two consecutive ones must lie an infinite number of zeroes. The simplest model is given on a "page" $Z \times Z$. Each one must be on its own line. We can do this by making the (a, b) position a one if and only $b = 0$. We think of this $Z \times Z$ as an ordinary page. For each a the set of (a, b) form a line. Thus an ordinary page may be thought of as a Z-sequence of distinguished lines. The ordinary page provides the desired model. We define a distinguished page of type i to be one where where all $(a, 0)$ are ones, $(0, 0)$ and $(0, i)$ are ones, and all other positions are zeroes.

$$\cdots\ 0\,0\,0\,0\,0 \cdots$$

Fig. 10.1. Ordinary Line. Model for $p \ll n^{-1}$

$$\begin{array}{c} \vdots \\ \cdots\ 0\,0\,0\,1\,0\,0\,0 \cdots \\ \cdots\ 0\,0\,0\,1\,0\,0\,0 \cdots \\ \cdots\ 0\,0\,0\,1\,0\,0\,0 \cdots \\ \vdots \end{array}$$

Fig. 10.2. Ordinary Page. Model for $n^{-1} \ll p \ll n^{-1/2}$

$$\begin{array}{c} \vdots \\ \cdots\ 0\,1\,0\,0\,0 \cdots \\ \cdots\ 0\,1\,0\,1\,0 \cdots \\ \cdots\ 0\,1\,0\,0\,0 \cdots \\ \vdots \end{array} \qquad \begin{array}{c} \vdots \\ \cdots\ 0\,1\,0\,0\,0 \cdots \\ \cdots\ 0\,1\,0\,0\,1\,0 \cdots \\ \cdots\ 0\,1\,0\,0\,0 \cdots \\ \vdots \end{array} \qquad \begin{array}{c} \vdots \\ \cdots\ 0\,1\,0\,0\,0 \cdots \\ \cdots\ 0\,1\,1\,0\,0\,0 \cdots \\ \cdots\ 0\,1\,0\,0\,0 \cdots \\ \vdots \end{array}$$

Fig. 10.3. Segment of Ordinary Book with Distinguished Pages of types 2, 3, 1. Model for $n^{-1/2} \ll p \ll n^{-1/3}$

When $n^{-1/2} \ll p(n) \ll n^{-1/3}$, we have a model on $Z \times Z \times Z$. The spikes (two ones finitely far apart) have to be separated by infinitely many ones, themselves infinitely far apart. Call the set of $\mathbf{x} = (a, b, c)$ with a fixed a page a. We insist that every page a be a distinguished page and let $t(a)$ denote its type. We further insist that the sequence $t(a)$, $-\infty < a < +\infty$ contains every finite string from the nonnegative integers as a substring. Such a Z-sequence of distinguished pages we shall call an ordinary book. These ordinary books (note that unlike ordinary pages they are not uniquely determined) provide the desired model.

The pattern shown for $k = 1$ and $k = 2$ does continue, although other complications arise. For example, when $k = 3$ distinguished books can either have a page with one line having three ones or a page with two lines each having two ones. The model on $Z \times Z \times Z \times Z$ is a volume consisting of a Z-sequence of distinguished books. Basically, there is a model of the complete theory when $n^{-1/k} \ll p(n) \ll n^{-1/(k+1)}$ on the set Z^{k+1}. Each of these models is constructed using the models of the previous theory as building blocks. Distinguished books are used to build ordinary volumes, giving a model for $k = 3$. Distinguished volumes are used to build ordinary libraries ($k = 4$), etc.

11. Stronger Logics

What happens if we increase the power of the first order language of graphs? The examples of this chapter show that, in many cases, the Zero-One Laws no longer hold. We shall only consider the basic case $p = \frac{1}{2}$. Even there relatively simple expansions of the language allow for sentences with no limiting probability.

11.1 Ordered Graphs

The ordered graph language consists of the binary relations of equality, adjacency, and order. A model is a graph G on an ordered set, which we'll write as $\{1, \ldots, n\}$. We shall restrict our attention to the random graph $G(n,p)$ with $p = \frac{1}{2}$. (Strictly speaking, ordered graphs are not a stronger logic but rather a stronger language.) The more powerful ordered language allows us to speak specifically about vertices $1, 2$. Thus we may write the property A that $1, 2$ are adjacent:

$$\exists_{x,y}\big[(x < y) \wedge (z < y \rightarrow z = x) \wedge x \sim y\big]$$

Clearly

$$\lim_{n \rightarrow \infty} \Pr\left[G\left(n, \frac{1}{2}\right) \models A\right] = \frac{1}{2}$$

so that when $p(n) = \frac{1}{2}$ the Zero-One Law cannot hold for the ordered language. Here we first show the much stronger negative result of nonconvergence, similar to the results of Section 8.3.4. In Section 11.1.2 we give a property with nonconvergent probability that could be considered natural. In Section 11.1.3 we give a positive result of Shelah that the nonconvergence cannot oscillate too badly.

11.1.1 Arithmetization

We copy the ideas of Section 8.3.3. Our task is somewhat easier because we already have a built in ordering. Note that vertices $1, 2$ can be defined as above. We first define an existential second order predicate $\text{ARITH}^*(u)$ which has the interpretation that we can place an arithmetic structure on $\{1, \ldots, u\}$. We first require of the binary relation D

$$D(1,y) \text{ if and only if } y=2$$
$$\text{For } x \neq 1,\ D(x,y) \text{ if and only if}$$
$$\text{there exist } z,w \leq u \text{ with } x=z+1,\ y=w+2 \text{ and } D(z,w)$$

Now, for $x,y \leq u$, $D(x,y)$ holds if and only if $y=2x$. Similarly require

$$E(1,y) \text{ if and only if } y=2$$
$$\text{For } x \neq 1,\ E(x,y) \text{ if and only if}$$
$$\text{there exist } z,w \leq u \text{ with } x=z+1,\ D(w,y) \text{ and } E(z,w)$$

Now, for $x,y \leq u$, $E(x,y)$ holds if and only if $y=2^x$. Similarly require

$$T(1,y) \text{ if and only if } y=2$$
$$\text{For } x \neq 1,\ T(x,y) \text{ if and only if}$$
$$\text{there exist } z,w \leq u \text{ with } x=z+1,\ E(w,y) \text{ and } T(z,w)$$

Now, for $x,y \leq u$, $E(x,y)$ holds if and only if y is the tower function $y=T(x)$ given by Definition 2.2. We define $\text{ARITH}^*(u)$ to be the second order existential predicate that there exist D,E,T with these properties.

Now we wish to replace the second order quantification with a purely first order sentence. We replace $\exists_D$ by $\exists_{c,d}$. Each instance of $D(x,y)$ is replaced by

$$x \leq y \leq u \text{ and there exists } z,\ c \leq z \leq d$$
$$\text{adjacent to } x,y \text{ and no other } w \leq u.$$

Replace E,T in the same way, always using different variable symbols. Let $\text{ARITH}(u)$ denote the new, now wholly first order, predicate.

But what is the relationship between ARITH and ARITH^*? Let us say c,d represent a set B of unordered pairs on $\{1,\ldots,u\}$ if B is precisely those pairs $\{x,y\}$ for which there exist z, $c \leq z \leq d$, adjacent to x,y and no other $w \leq u$. We need a result similar to Representation Theorem 8.1.1.

Theorem 11.1.1. *Set $u = \ln^{1/4} n$. Almost surely for every set B of unordered pairs on $\{1,\ldots,u\}$ there exist c,d representing B.*

Proof: Split $u+1,\ldots,n$ into $\sim nu^{-2}$ disjoint intervals $[c,d]$ of size u^2. A given interval represents a given B with probability at least 2^{-u^3} as there is at least one way that it can represent it. These events are independent over the different intervals so the probability that no interval represents a given B is less than $(1-2^{-u^3})^{nu^{-2}} \leq e^{-s}$ where we set $s = 2^{-u^3}nu^{-2}$ for convenience. Our choice of u (not best possible) assures us that s is dominated by the factor n, indeed that $s = n^{1-o(1)}$. There are less than 2^{u^2} different B so the probability that some B is not represented is less than $2^{u^2}e^{-s}$. Our choice of u now assures us that this product is dominated by the e^{-s} factor. In particular, it approaches zero.

Since $\text{ARITH}^*(u)$ is true for all u (we can put arithmetic on the first u integers) Theorem 11.1.1 gives that $\text{ARITH}(u)$ is almost surely true for $u = \ln^{1/4} n$. We extend to

$$\text{MaxArith}(u)\colon \text{Arith}(u) \wedge \neg\exists_y \text{Arith}(y) \wedge u < y$$

that u is the largest element so that $\{1, \ldots, u\}$ can be so arithmeticized. We now define $v = \log^* u$ by $\text{MaxArith}(u)$ and $\exists_{x \le u} T(v, x)$ and $\neg\exists_{x \le u} T(v + 1, x)$ where T is as previously defined. As $\ln^{1/4} n \le u \le n$ (normally a huge range but not here!) $\log^* n - 2 \le v \le \log^* n$. Our sentence A is that $v \equiv 0, 1, 2$ or 3 modulo 8. This is easy to write:

$$\exists_{a,b,c,d \le u} D(a, b) \wedge D(b, c) \wedge D(c, d) \wedge [u = d \vee u = d + 1 \vee u = d + 2 \vee u = d + 3]$$

When $\log^* n \equiv 7 \pmod 8$ this almost surely fails while when $\log^* n \equiv 3 \pmod 8$ this almost surely holds.

11.1.2 Dance Marathon

Imagine n couples at a Dance Marathon. Each dance each couple remains standing with independent probability one half. A couple that does not remain standing is removed from the competition. A couple wins the prize if at the end of some dance they are the only couple that remained standing. It may happen that none of the couples that began a dance remained standing, in which case the prize is not awarded. What is the probability $f(n)$ that the prize is awarded?

The surprising fact is that $\lim_{n \to \infty} f(n)$ does not exist. We have an exact formula

$$f(n) = n \sum_{k=1}^{\infty} (1 - 2^{-k})^{n-1} 2^{-k-1}$$

Imagine the marathon continuing until the winning couple also collapses and let k be the number of dances completed by the winners. There are n choices for the winning couple, k can be any positive integer, the other $n - 1$ couples all do not survive the first k dances, and the winning couple survives the first k dances and not the $k+1$-st dance. Now parametrize $n = 2^u \theta$ with u integral and $\theta \in [1, 2)$. For $k = u + i$, with i fixed and $u \to \infty$, $n(1 - 2^{-k})^{n-1} 2^{-k-1} \sim \theta 2^{-i} e^{-2^i \theta}$. Set

$$g(\theta) = \sum_{i=-\infty}^{+\infty} \theta 2^{-i} e^{-2^i \theta}$$

Some careful analysis gives that for θ fixed and $n = 2^u \theta \to \infty$, $f(n) \to g(\theta)$ and simple calculation gives that $g(\theta)$ is not a constant.

This yields a property Dm in the ordered graph language with no limit probability which some may actually consider natural: There exists a k with precisely one y adjacent to $1, \ldots, k$. Formally:

$$\text{Dm}\colon \exists_k \exists_y \forall_u \Big(y = u \leftrightarrow \big[\forall_z z \le k \to z \sim u\big]\Big)$$

Each y is like a couple, adjacency with $i = 1, 2, 3, \ldots$ meaning they survived the first, second, third,... dance. There is a minor technical glitch in that y cannot be adjacent to y but this is asymptotically insignificant and when $n = 2^u\theta \to \infty$ with $\theta \in [1,2)$ fixed $\Pr[\text{DM}] \to g(\theta)$ so that $\lim_{n\to\infty} \Pr[\text{DM}]$ does not exist.

11.1.3 Slow Oscillation

We begin with a result that takes us into the realm of circuit complexity, whose argument we only indicate.

Theorem 11.1.2. *Let A be a sentence in the ordered graph language and let H be an arbitrary graph on vertex set $V = \{1,\ldots,2n+1\}$. Let $g_H(i)$, i=n,n+1, be the probability that $H|_S$ satisfies A when S is a uniformly chosen subset of $V(H)$ of size i. Then $g_H(n+1) - g_H(n) \to 0$. More precisely, fixing A there is a function $\epsilon(n) \to 0$ so that, for any H of size $2n+1$, $\left|g_H(n+1) - g_H(n)\right| < \epsilon(n)$.*

With A, H fixed we may think of the event $H|_S \models A$ as a Boolean function of S. To make this transformation express A in prenex normal form (all quantifiers in front) and replace each $\exists_x P(x)$ by the $\vee$ over $1 \le i \le 2n+1$ of $(i \in S) \wedge P(i)$ and each $\forall_y Q(y)$ by the $\wedge$ over $1 \le i \le 2n+1$ of $(i \in S) \Rightarrow Q(y)$ This reduces A to a Boolean combination of statements of the forms $i \in S$, $i < j$, $i = j$ and $i \sim j$. All but the first have a definite truth value, given H. Thus $H|_S \models A$ becomes a Boolean expression in the atomic variables $i \in S$. In terms of circuit complexity each existential quantifier has become an OR gate with fan-in $2n+1$ and each universal quantifier has become an AND gate with fan-in $2n+1$. The depth of the circuit is simply the quantifier depth of A. We then have a bounded depth polynomial size circuit. A deep theorem in circuit complexity is that such a circuit cannot compute the Boolean function majority – in our case that more than half of the $i \in S$ atomic variables are true. Moreover, let $h(n), h(n+1)$ denote the probabilities that the circuit accepts an input with a random n or $n+1$ of the atomic variables set true. With further technical work it can be shown that $h(n+1) - h(n) \to 0$ in the appropriate uniform sense, giving Theorem 11.1.2.

Theorem 11.1.3 (Shelah). *For any A in the ordered language, letting $f(n)$ denote the probability $G(n, \frac{1}{2})$ models A,*

$$\lim_{n\to\infty} f(n+1) - f(n) = 0$$

Proof: We may generate a random graph on n vertices by first generating a random graph H on $2n+1$ vertices and then taking the induced subgraph of H on a randomly selected n vertices, preserving the original order. Thus, by this roundabout way,

$$f(n) = \sum_H \Pr[H] g_H(n)$$

where H ranges over all labelled graphs on $1, \ldots, 2n+1$, $\Pr[H]$ is the probability that the random graph is H, and $g_H(n)$ is given by Theorem 11.1.2 Similarly,

$$f(n+1) = \sum_H \Pr[H] g_H(n+1)$$

so that

$$\left| f(n+1) - f(n) \right| \leq \sum_H \Pr[H] \left| g_H(n+1) - g_H(n) \right|$$

But $\sum_H \Pr[H] = 1$ so $\left| f(n+1) - f(n) \right| \leq \max_H \left| g_H(n+1) - g_H(n) \right|$ which goes to zero.

One powerful consequence is that in the ordered language one cannot probabilistically approximate the property that the number of vertices is even. That is, there is no sentence in the ordered language so that A holds almost surely when the number of vertices is even and almost never when the number of vertices is odd. Negative results on approximate representations are rare gems indeed.

11.2 Existential Monadic

Here we remove the ordering of Section 11.1 but consider sentences in a stronger logic. We restrict our attention to $G(n,p)$ with $p = \frac{1}{2}$. It is not particularly difficult to show that second order logic does not have convergence but it is perhaps surprising how little of second order logic we need to destroy convergence.

Definition 11.1 *A property is said to* existential monadic *if it is of the form*

$$\exists_{U_1} \cdots \exists_{U_r} P$$

where the initial existential quantifiers (and they must all be existential and not universal) are over unary predicates and P is a first order sentence in equality, adjacency, and the unary U_i.

It will be convenient to use the association between unary predicates and sets and consider the sentence of the form: "The exist sets $S_1, \ldots, S_r$ such that ...," where we allow statements $x \in S_i$ in the remaining portion.

Theorem 11.2.1. *There is an existential monadic A for which*

$$\lim_{n \to \infty} \Pr\left[G\left(n, \frac{1}{2}\right) \models A \right]$$

does not exist.

We recall the second order monadic sentence BIGGAP of Section 8.3.3. It had the property that it held on a set S of size s if and only if $\log^* s$ modulo 100 is one of $1, \ldots, 50$. Applying Claim 8.2.2 we create a first order sentence BIGGAP*. It has two properties:

1. If BIGGAP* holds for some finite graph G on t elements then BIGGAP holds for some subgraph $H = G|_S$ with S having s elements and $s \leq t = O(s^2)$.
2. If BIGGAP holds for some finite graph H on an s element set S then BIGGAP* holds for some extension G on t elements with $s \leq t = O(s^2)$.

As squaring can only change $\log^* s$ (for s sufficiently large) by at most one we deduce that if $\log^* t$ modulo 100 is one of $52, \ldots, 99$ modulo 100 then BIGGAP* holds for no graph of size t.

Now we give the sentence A promised by Theorem 11.2.1

A: There exist sets $T = T_1, T_2, T_3, T_4, P$ such that

1. The sets are disjoint.
2. Adjacency gives a bijection on each $T_i \times T_j$. That is, for each $i \neq j$,
$$\forall_{x \in T_i} \exists!_{y \in T_j} x \sim y$$
3. P is in bijective correspondence with $T_1 \times T_2 \times T_3 \times T_4$:
$$\forall_{x_1 \in T_1, x_2 \in T_2, x_3 \in T_3, x_4 \in T_4} \exists!_{y \in P} (y \sim x_1 \wedge y \sim x_2 \wedge y \sim x_3 \wedge y \sim x_4)$$
and
$$\forall_{y \in P} \exists!_{x_1 \in T_1, x_2 \in T_2, x_3 \in T_3, x_4 \in T_4} (y \sim x_1 \wedge y \sim x_2 \wedge y \sim x_3 \wedge y \sim x_4)$$
4. No two vertices of the graph outside $P \cup T_1 \cup \ldots \cup T_4$ have the same adjacency pattern to P. That is, for all $x \neq y \notin P \cup T_1 \cup \ldots \cup T_4$ there exists $z \in P$ with either $x \sim z$ and $\neg y \sim z$ or $y \sim z$ and $\neg x \sim z$.
5. The property BIGGAP* holds for the graph restricted to T.

First suppose $\log^* n$ is one of $54, \ldots, 99$ modulo 100. Set $|T| = t$. From the fourth property $|P| \geq \log_2 n$ and from the first two properties $|P| = t^4$ so that $t \geq \log_2^{1/4} n$. Thus $\log^* t$ is one of $52, \ldots, 99$ modulo 100 and so BIGGAP* cannot hold on the graph restricted to T. Thus $\Pr[A] \to 0$ on this subsequence.

The positive existence of these sets first uses a natural extension of the Alice's Restaurant property shown to hold almost always in Claim 0.1.3.

Claim 11.2.2 *Let $\epsilon > 0$ be fixed and suppose $k = k(n) < (1 - \epsilon) \log_2 n$. Then almost surely $G \sim G(n, \frac{1}{2})$ has the k-Alice's Restaurant property of Definition 2.3.*

Proof: There are $\binom{n}{k}2^k \leq n^k$ choices of disjoint sets X, Y of vertices with $|X \cup Y| = k$. As in Claim 0.1.4 the probability that there is no witness z adjacent to all the $x \in X$ and none of the $y \in Y$ is $(1-2^{-k})^{n-k}$. The probability of the k-Alice's Restaurant property failing is then less than $n^k(1-2^{-k})^{n-k}$ which goes to zero for this k.

Now suppose $\log^* n$ is one of $5, \ldots, 49$ modulo 100. Set $s = 10\ln^{1/4} n$. We know there is some graph H on t vertices satisfying BIGGAP* with $s \leq t \leq O(s^2)$. We apply Claim 11.2.2 repeatedly, beginning with the empty set, to find a copy of H, let $T = T_1$ be its vertex set. We continue applying Claim 11.2.2 to get the desired graph on $T_1 \cup T_2 \cup T_3 \cup T_4$. Now apply Claim 11.2.2 with $k = 4t$ to find, for each $t_1 \in T_1, \ldots, t_4 \in T_4$ a y adjacent to just those t_i in $T_1 \cup T_2 \cup T_3 \cup T_4$. Let P be the set of those y. Observe that in the creation of the T_i and P we only looked at the potential edges to the vertices already specified. Hence after having created the T_i and P we may still regard the remaining pairs as being adjacent with independent probability $\frac{1}{2}$. In particular, for each $x, y \notin P \cup T_1 \ldots \cup T_4$ the adjacencies to P are independent. The probability that they have the same adjacency pattern is thus $2^{-|P|}$. Our lower bound on s insures that $|P| \geq 10^4 \log_2 n$ so that this probability is $o(n^{-2})$ and almost surely no x, y have the same adjacencies. Hence the sentence A holds with probability approaching one under this restriction. Thus the limit probability of A does not exist, completing the proof of Theorem 11.2.1.

12. Three Final Examples

We close with three quite different random structures. While the structures are new the techniques we have already developed allow us a rather complete analysis of their First Order worlds. We wish our readers equal success.

12.1 Random Functions

Let $f\colon \{1,\dots,n\} \rightarrow \{1,\dots,n\}$ be uniformly chosen from the n^n possible such functions. We write $x \rightarrow y$ for $f(x) = y$. Let F_n denote this random model. We consider the first order language, denoted FUNC, with equality and the binary relation $\rightarrow$. There is a single axiom $\forall_x \exists!_y x \rightarrow y$. (This guarantees that $\rightarrow$ represents a well defined function. Alternately, we could speak of a first order theory with a single function symbol F.) The study of this system has been made in some detail by Lynch [13], here we outline the results. Is there a Zero-One Law? No! Consider the statement

$$\text{NFP:} \qquad \neg\exists_x x \rightarrow x$$

which says that f has no fixed point. For each x the probability that $f(x) = x$ is $\frac{1}{n}$ and these events are independent. Thus $\Pr\left[F_n \models \text{NFP}\right] = \left(1 - \frac{1}{n}\right)^n$ which has limiting value e^{-1}. More generally we say distinct $x_1, \dots, x_i$ form an i-cycle if $x_1 \rightarrow x_2 \rightarrow \dots \rightarrow x_i \rightarrow x_1$, fixed points being 1-cycles. There are $(n)_i/i$ potential i-cycles (as any of the i values could be made the initial value), each with probability n^{-i} of being an i-cycle. The expected number of i-cycles then has limiting value $\frac{1}{i}$ and one can show that the number X_i of i-cycles has limiting distribution Poisson with mean $\frac{1}{i}$.

The analysis of F_n bears a striking resemblance to that of the random graph $G(n,p)$ with $p = \frac{1}{n}$ as described in Section 3.7. The key is to consider y a child of x if $y \rightarrow x$ and $y \neq x$. Each x has $n-1$ potential children and each is a child with independent probability $\frac{1}{n}$. Hence the number of children of x has, asymptotically, a Poisson distribution with mean one. Let T_v be the tree generated from v by this process. The limiting distribution of T_v is that given by Poisson childbirth with mean one. Such trees are finite with probability one although the analysis is delicate – indeed, the expected size of T_v is infinite.

It is helpful here to think of the language FUNC augmented by a constant symbol v. First order sentences in the language of rooted trees could then be translated into the augmented FUNC. As in Section 3.7.1 we translate our favorite: Eve has no children that have no children which have no children. This becomes

$$\neg\exists_w\Big(w \to v \wedge \neg\exists_x\big(x \to w \wedge (\neg\exists_y y \to x)\big)\Big)$$

The language FUNC does not actually contain constant symbols. Basically, those x lying in small cycles are special and treated like constant symbols. For each $i \geq 1$ the number of i-cycles has Poisson distribution with mean i^{-1}. From each $x = x_j$ of such an i-cycle let T_x^* be the tree generated from x as described above except that x_{j-1} is not counted as a child of x_j. The T_x^* then have limiting distribution given by Poisson childbirth with mean one.

Let s be an arbitrary but fixed positive integer. We wish to characterize the limit probabilities of first order sentences A with quantifier depth at most s. Let $x_1, \ldots, x_i$ be an i-cycle. For each x_j let β_j denote the $(3^s, s)$ value (as defined in Section 3.3.1) of $T_{x_j}^*$. We call $\beta = (\beta_1, \ldots, \beta_i)$ the s-type of the i-cycle, with the proviso that it should be independent of the particular labelling – for example, with $k = 4$, (A, B, C, D), (B, C, D, A), (C, D, A, B), (D, A, B, C) (but not (D, C, B, A)) are all regarded the same. Let Ω_s denote the set of s-types of i-cycles for $1 \leq i \leq 2\cdot 3^s$. Let $g \colon \Omega_s \to \{0, 1, \ldots, s-1, M\}$ where M denotes "many." Let L_g be the first order statement that for each $\beta \in \Omega_s$ there are precisely $g(\beta)$ i-cycles of type β, or at least s i-cycles of type β when $g(\beta) = M$. The number of i-cycles of type β has a Poisson distribution with mean $\frac{B}{i}$ where B is the probability that i independent Poisson childbirths with mean one have $(s, 3^s)$ types in cyclic order β. We have seen in Section 3.7.1 that B is the sum and product of terms generated from the rationals by addition, subtraction, multiplication and base e exponentiation.

Let T be a finite rooted tree. We naturally associate with T a first order sentence A_T in the language FUNC with the meaning that for some x the tree generated from x is isomorphic to T. (For example, if T consists of a root r and three children then A_T is that there exist distinct x, y_1, y_2, y_3 with $y_1 \to x$, $y_2 \to x$, $y_3 \to x$ and no z with $z \to y_1$ or $z \to y_2$ or $z \to y_3$ and no z other than y_1, y_2, y_3 with $z \to x$.) It can be shown that each A_T holds almost surely. Let TH be the almost sure theory of F_n, so that it contains all these sentences A_T. Each L_g gives an s-completion of this theory as one can show combinatorially that the Duplicator can win s-move Ehrenfeucht game on any f, f' satisfying TH $+ L_g$. Thus for any sentence A

$$\lim_{n\to\infty} \Pr\big[F_n \models A\big] = \sum \lim_{n\to\infty} \Pr\big[F_n \models L_g\big]$$

with the sum over those g such that TH $+ L_g \models A$. We deduce:

Theorem 12.1.1. *For any sentence A in* FUNC

$$\lim_{n\to\infty} \Pr\big[F_n \models A\big]$$

exists and can be generated from the rational numbers through addition, subtraction, multiplication and base e exponentiation.

Examples: Let A be the sentence that there are no fixed points which have no children that have no children. Explicitly:

$$\neg\exists_x\big(x \to x \wedge \neg\exists_y(y \to x \wedge \neg\exists_z z \to y)\big)$$

In Poisson childbirth with mean one the probability that there are no children with no children is $\exp[-e^{-1}]$. Thus the number of fixed points with no children with no children has Poisson distribution with mean $\exp[-e^{-1}]$ and so that limiting probability of A is $\exp\big[-\exp[-e^{-1}]\big]$. Let B be the statement that there is an isolated 2=cycle. Explicitly

$$\exists_x\exists_y\big[(x \to y) \wedge (y \to x) \wedge \neg\exists_z(z \to x) \wedge \neg\exists_z(z \to y)\big]$$

Given x, y the probability that neither has a child is e^{-2}. The expected number of 2-cycles is $\frac{1}{2}$ and the expected number of isolated 2-cycles is $\frac{1}{2}e^{-2}$. The number of isolated 2-cycles has a Poisson distribution and the probability that there is (at least) one is one minus the probability that there are none. Hence B has limiting probability $1 - \exp\big[-\frac{1}{2}e^{-2}\big]$.

12.2 Distance Random $G(n, \overline{p})$

Let $\overline{p} = (p_1, p_2, \ldots)$ be an infinite sequence with all $p_i \in [0, 1]$. Łuczak and Shelah [12] have examined a random structure they denoted $G(n, \overline{p})$. This is a graph with vertices $1, \ldots, n$. For each pair of distinct vertices i, j

$$\Pr\big[\{i, j\} \in G\big] = p_{|j-i|}$$

and these events are independent over all pairs. That is, the probability of adjacency depends on the "distance" $|j - i|$ between the vertices. As with the usual $G(n, p)$ there are a variety of different ranges of the sequences $\overline{p}$. We shall examine $G(n, \overline{p})$ both from the vantage point of the ordinary first order language on graphs (with just equality and adjacency) and the ordered graph language of Section 11.1 which includes less than.

12.2.1 Without Order

Here we examine $G(n, \overline{p})$ from the vantage point of the ordinary first order language on graphs, with equality and adjacency but without order. We shall assume that $0 < p_i \leq \frac{1}{2}$ for all i (to avoid technical difficulties) and that $\sum_{i=1}^{n} p_i = o(\ln n)$. The second assumption makes our random graph, in some sense, rather sparse. Our goal is Theorem 12.2.5, a Zero-One Law for $G(n, \overline{p})$.

Definition 12.2.1. *Let* $H_1 = (V(H_1), E(H_1))$, $H_2 = (V(H_2), E(H_2))$ *be graphs on disjoint vertex sets. Their* sum $G = H_1 + H_2$ *is given by* $G = (V(H_1) \cup V(H_2), E(H_1) \cup E(H_2))$. *That is,* G *is the disjoint union of* H_1 *and* H_2.

Now we emulate Theorem 10.2.2.

Theorem 12.2.1. *If* $H_1 \equiv_k H_1'$ *and* $H_2 \equiv_k H_2'$ *then* $H_1 + H_2 \equiv_k H_1' + H_2'$.

Proof: Consider the game on the two models $H_1 + H_2$ and $H_1' + H_2'$. When Spoiler selects an $x \in H_1$ Duplicator pretends that she is playing the k-move game on H_1 and H_1' and makes the corresponding reply on H_1'. When Spoiler selects an $x \in H_2$ Duplicator pretends that she is playing the k-move game on H_2 and H_2' and makes the corresponding reply on H_2'. Plays on H_1' and H_2' are treated similarly. Duplicator is then preserving the adjacencies inside the submodels. But if $x \in H_1$ and $y \in H_2$ have been selected then their analogous vertices also have $x' \in H_1'$ and $y' \in H_2'$. Since *every* $x \in H_1$ is nonadjacent to every $y \in H_2$ (and also for H_1', H_2') Duplicator has preserved adjacency in these cases as well.

Definition 12.2.2. *The* Ehrenfeucht Semigroup $\text{EHR}[k]$ *for graphs is the set of* k*-equivalence classes of graphs* H *under the induced operation of* $+$ *as defined above.*

As in Section 10.2.2 $\text{EHR}[k]$ is a finite set. Here, unlike with strings, the operation $+$ is Abelian and so $\text{EHR}[k]$ forms an Abelian semigroup. For s a positive integer, $w \in \text{EHR}[k]$ we naturally let sw denote the sum of w with itself s times. Analogous to Theorem 10.2.4 we have a somewhat simpler result.

Theorem 12.2.2. *If* $s, t \geq k$ *then* $sw = tw$.

Proof: Consider the Ehrenfeucht Game with one graph consisting of s copies of H and the other consisting of t copies of H. Each time Spoiler moves to a new copy of H Duplicator can do the same, as the game lasts only k rounds. Moves inside a copy of H are responded to with the same move in a corresponding copy.

Theorem 12.2.3. *There exists an* $w \in \text{EHR}[k]$ *such that* $w + v = w$ *for all* $v \in \text{EHR}[k]$.

Proof: Let w be the sum of kv over all $v \in \text{EHR}[k]$. For any particular v we may write $w = kv + w'$ so that $w + v = (k+1)v + w' = kv + w' = w$.

Theorem 12.2.4. *Let* $\overline{p}$ *satisfy* $\sum_{i=1}^{n} p_i = o(\ln n)$ *and* $0 < p_i \leq \frac{1}{2}$ *for all* i. *Let* H *be any fixed graph. Then almost surely* H *is an isolated subgraph of* $G(n, \overline{p})$.

We outline the argument. Label the vertices of H by $1, \ldots, s$. Let $G \sim G(n, \overline{p})$. For $0 \leq u \leq \frac{n-s}{s}$ let B_u be the event that G restricted to $us + 1, \ldots, us + s$ is isomorphic to H under the mapping $x \rightarrow us + x$ and that there are no edges from $us+1, \ldots, us+s$ to any other of the vertices $1, \ldots, n$. Let X denote the number of B_u that hold. Fix u. There is a constant positive probability c that G restricted to $us+1, \ldots, us+s$ is isomorphic to H under the mapping $x \rightarrow us + x$. The probability that there are no edges from $us + 1, \ldots, us + s$ to any other of the vertices $1, \ldots, n$ is at least $\left[\prod_{i=1}^{n}(1-p_i)\right]^{2s}$. As $1 - p_i > \exp[-10p_i]$ and as $\sum p_i = o(\ln n)$ this probability is $n^{-o(1)}$. Hence the expectation $E[X]$ is $n^{1-o(1)}$, and so goes to infinity. Showing that $X > 0$ almost surely requires a somewhat technical application of the Second Moment Method, and is given as Lemma 10.30 of [9].

Theorem 12.2.5. *Let $\overline{p}$ satisfy $\sum_{i=1}^{n} p_i = o(\ln n)$ and $0 < p_i \leq \frac{1}{2}$ for all i. For any sentence A in the first order theory of graphs (without order)*

$$\lim_{n\rightarrow\infty} \Pr\left[G(n, \overline{p}) \models A\right] = 0 \text{ or } 1$$

Proof: Let A have quantifier depth k. From Theorem 12.2.3 there exists H so that all graphs G containing H as an isolated subgraph have the same k-Ehrenfeucht value and hence either all satisfy A or all do not satisfy A. From Theorem 12.2.4 $G(n, \overline{p})$ almost surely does contain H as an isolated subgraph.

12.2.2 With Order

Here we examine $G(n, \overline{p})$ from the vantage point of the ordered first order language on graphs, with equality and adjacency and order. We shall assume that $0 < p_i \leq \frac{1}{2}$ for all i (to avoid technical difficulties) and that $\sum_i ip_i$ converges. The second assumption makes our random graph quite sparse, more so than in the previous section. In the ordered graph language we can say that vertices 1 and 2 are adjacent. Hence there is no hope for a Zero-One Law. Rather, our goal is Theorem 12.2.10, that all limit probabilities do exist. Our models will be ordered graphs, which can be written $G = \left(V(G), E(G), <_G\right)$, consisting of a vertex set, an edge set, and an ordering. We first modify Definition 12.2.1 to give a sum of these structures.

Definition 12.2.3. *Let $H_1{=}\left(V(H_1), E(H_1), <_1\right)$, $H_2{=}\left(V(H_2), E(H_2), <_2\right)$ be graphs on disjoint ordered vertex sets. Their* ordered sum $G = H_1 \oplus H_2$ *is the ordered graph with $G = \left(V(H_1) \cup V(H_2), E(H_1) \cup E(H_2), <\right)$. Here $<$ is given by setting $h_1 < h_2$ for all $h_1 \in V(H_1), h_2 \in V(H_2)$ and using $<_1$ when $h_1, h_2 \in V(H_1)$ and $<_2$ when $h_1, h_2 \in V(H_2)$.*

Less formally, $H_1 \oplus H_2$ is given by placing H_1 to the left of H_2. Analogously to Chapter 2 and Section 10.2.1 we may define the k-move Ehrenfeucht game

EHR$(G_1, G_2; k)$ between Spoiler and Duplicator on ordered graphs G_1, G_2. In addition to the usual requirements on graphs Duplicator must here insure that order is preserved: if x_i, x_j were selected in G_1 and y_i, y_j we the vertices of G_2 selected on the same rounds then $x_i < x_j$ if and only if $y_i < y_j$. We then define $G_1 \equiv_k G_2$ if Duplicator wins this game EHR(G_1, G_2, k). As with Theorem 2.3.1 for graphs without order, $G_1 \equiv_k G_2$ if and only if they have the same truth value for all first order sentences A of quantifier depth at most k.

Now we emulate Theorems 10.2.2 and 12.2.1 and Definitions 10.2 and 12.2.2 for ordered graphs.

Theorem 12.2.6. *If $H_1 \equiv_k H_1'$ and $H_2 \equiv_k H_2'$ then $H_1 \oplus H_2 \equiv_k H_1' \oplus H_2'$.*

Proof: Consider the game on the two models $H_1 \oplus H_2$ and $H_1' \oplus H_2'$. When Spoiler selects an $x \in H_1$ Duplicator pretends that she is playing the k-move game on H_1 and H_1' and makes the corresponding reply on H_1'. When Spoiler selects an $x \in H_2$ Duplicator pretends that she is playing the k-move game on H_2 and H_2' and makes the corresponding reply on H_2'. Plays on H_1' and H_2' are treated similarly. Duplicator is then preserving the adjacencies and order inside the submodels. But if $x \in H_1$ and $y \in H_2$ have been selected then their analogous vertices also have $x' \in H_1'$ and $y' \in H_2'$. Since *every* $x \in H_1$ is nonadjacent to every $y \in H_2$ (and also for H_1', H_2') Duplicator has preserved adjacency in these cases as well. Since *every* $x \in H_1$ is less than every $y \in H_2$ (and also for H_1', H_2') Duplicator has preserved order in these cases as well.

Definition 12.2.4. *The* Ehrenfeucht Semigroup EHR$[k]$ *for ordered graphs is the set of k-equivalence classes of graphs H under the induced operation of $\oplus$ as defined above.*

As before EHR$[k]$ is a finite semigroup, though unlike the semigroup for graphs under $+$ given by Definition 12.2.2 it is in general not Abelian. For $w \in$ EHR$[k]$, s a positive integer we naturally define sw as the sum, under $\oplus$, of s copies of w. We do have the same result as Theorem 10.2.4 on strings.

Theorem 12.2.7. *Let $w \in$ EHR$[k]$, with the Ehrenfeucht Semigroup for ordered graphs as given in Definition 12.2.4. Let $s, t \geq 2^k - 1$. Then $sw = tw$.*

Proof: Let G be an ordered graph in class w and let sG denote the sum, under $\oplus$, of s copies of G. It suffices to give a Duplicator strategy for EHR$[sG, tG, k]$. The proof of Theorem 10.2.4 gives a strategy for Duplicator in the k-move Ehrenfeucht game on strings $s\alpha, t\alpha$. Now in EHR$[sG, tG, k]$ suppose Spoiler selects some v in the i-th copy of G in sG. Duplicator thinks of Spoiler has having selected the i-th α in $s\alpha$, for which she has a response the jth α in $t\alpha$. She then goes to the j-th copy of G in tG and selects the identical vertex. Moves by Duplicator in the same copy of G in sG are responded to by playing the identical vertices in the same copy of G in tG and so adjacency and order are preserved. With moves in different copies of G Duplicator survives as she has preserved order.

Theorem 12.2.8. *Let k be an arbitrary positive integer. Then there exists a finite ordered graph H such that*

$$P \oplus H \oplus M \oplus H \oplus S \equiv_k P \oplus H \oplus M' \oplus H \oplus S$$

for all ordered graphs P, S, M, M'.

Proof: The set of k-equivalence classes of finite ordered graphs forms a semigroup with identity (the pointless graph) that satisfies Theorem 12.2.7. We may therefore apply the results of Section 10.3 to this semigroup. In particular, from Property 10.3.8 there exists a persistent w. Let H be an ordered graph in class w. From Property 10.3.7 H has the desired property.

Theorem 12.2.9. *Let $\overline{p}$ be such that $\sum_i ip_i$ converges and $0 < p_i \leq \frac{1}{2}$ for all i. Let H be any fixed graph on labelled vertices $1, \ldots, s$. Let $G \sim G(n, \overline{p})$. Let $0 \leq \alpha < \beta \leq 1$ be fixed. Then almost surely there exists u with $n\alpha < u < n\beta$ such that*

1. *G restricted to $u+1, \ldots, u+s$ is H by the isomorphism $x \rightarrow x+u$.*
2. *G has no edges $\{x, y\}$ with $x \in \{u+1, \ldots, u+s\}$ and $y \notin \{u+1, \ldots, u+s\}$.*
3. *G has no edges $\{x, y\}$ with $x \leq u$ and $u+s < y$.*

Note that such graphs may be written $G = P \oplus H \oplus S$. We outline the argument. For each u let B_u be the event that the above holds. There is a positive constant probability c (dependent on H and $\overline{p}$) that G restricted to $u+1, \ldots, u+s$ is H as described. For $1 \leq k \leq n$ there are at most k pairs x, y with $y - x = k$ and $x \leq u+s < y$ and at most k pairs x, y with $y - x = k$ and $x \leq u < y$. The probability of no edges of the second or third types is then at least $\prod_{k=1}^{n}(1-p_k)^{2k}$. As all $1 - p_k \geq \exp[-10p_k]$ and $20\sum_{k=1}^{n} kp_k$ is bounded by a constant independent of n (as, critically, $\sum kp_k$ converges) we bound $\Pr[B_u]$ from below by a positive constant independent of n. Let X be the number of u with $n\alpha < u < n\beta$ such that B_u holds. Then $E[X]$ approaches infinity. We omit the technical work required to show that X is almost surely nonzero.

Theorem 12.2.10. *Let $\overline{p}$ be such that $\sum ip_i$ converges and $0 < p_i \leq \frac{1}{2}$ for all i. For any sentence A in the first order theory of ordered graphs*

$$\lim_{n\rightarrow\infty} \Pr\big[G(n, \overline{p}] \models A\big]$$

exists.

Proof: We fix A with quantifier depth k, $\overline{p}$ satisfying the conditions, and set $\mu_n = \Pr\left[G(n,\overline{p}) \models A\right]$ for notational convenience. Fix an H with the properties of Theorem 12.2.8. Let $n < m$ where we shall be interested in the case n large. We jointly generate $G(n,\overline{p})$, $G(m,\overline{p})$ as follows. First we generate $G_n \sim G(n,\overline{p})$. Let $i,j \in \{1,\ldots,m\}$. When $i,j \leq \frac{n}{2}$ we make i,j adjacent in G_m if and only if they are adjacent in G_n. When $i,j \geq m-\frac{n}{2}$ we make i,j adjacent in G_m if and only if $i-(m-n), j-(m-n)$ are adjacent in G_n. Otherwise i,j are adjacent with independent probability $p_{|j-i|}$. (That is, the left and right halves of G_n become the left and right sides of G_m, a random middle part is added with the other adjacencies of G_m decided by the usual probabilities.) Fix ϵ positive. By Theorem 12.2.9 we select n sufficiently large so that with probability at least $1-\frac{\epsilon}{2}$ G_n has copies of H in both the regions $\left[\frac{n}{6},\frac{n}{3}\right]$ and $\left[\frac{2n}{3},\frac{5n}{6}\right]$, i. e. , near the center of each half. Further, these H copies are separating in the sense of Theorem 12.2.9. We may then write $G_n = P \oplus H \oplus M \oplus H \oplus S$. Now consider the possible adjacencies x,y of G_m where $\frac{n}{2} < x < m-\frac{n}{2}$ and either $y < \frac{n}{3}$ or $y > m-\frac{n}{3}$. This includes all possible adjacencies between the middle part of G_m and vertices to the left of the left H or to the right of the right H. There are at most $\frac{2n}{3}$ pairs x,y with any fixed $|y-x|$ and we always have $|y-x| \geq \frac{n}{6}$. The probability that there are any adjacencies is at most the expected number of such adjacencies which is at most

$$\frac{2n}{3}\sum_{k\geq\frac{n}{6}} p_k \leq 4\sum_{k\geq\frac{n}{6}} kp_k \leq \frac{\epsilon}{2}$$

for n sufficiently large. Here we have made critical use of the convergence of $\sum_k kp_k$.

With probability at least $1-\epsilon$ we may write $G_n = P \oplus H \oplus M \oplus H \oplus S$ and $G_m = P \oplus H \oplus M' \oplus H \oplus S$ with the same P, S. From Theorem 12.2.8 it follows that G_n, G_m have the same truth value for A. Hence $|\mu_n - \mu_m| \leq \epsilon$. But ϵ was arbitrarily small. We have shown that the sequence μ_n forms a Cauchy sequence. Hence the sequence μ_n converges.

12.3 Random Lifts

Let H be a fixed graph. Let n be a positive integer. The study of the random lift $L_n(H)$ was initiated by Nati Linial.

Definition 12.3.1. *The* random lift *$L_n(H)$ is a graph whose vertices are labelled (v,i) where v is a vertex of H and $1 \leq i \leq n$. The vertices (v,i) for a fixed v are called the* fibre *of v. These vertices are mutually nonadjacent. When v,w are nonadjacent vertices in H there are no edges between their respective fibres. When v,w are adjacent the edges between their fibres are given by a randomly chosen* 1*-factor. The choices of* 1*-factors are mutually independent.*

Equivalently, for all adjacent $v, w \in H$ permutations $\sigma_{vw} \in S_n$ are selected uniformly and independently (selecting only one of σ_{vw}, σ_{wv}) and (v, i) is adjacent to $\big(w, \sigma_{vw}(i)\big)$. We shall say that the points of the fibre of v lie over v.

For given n, H and property A we can naturally speak of $\Pr\big[L_n(H) \models A\big]$. Our concern will be the limits of such probabilities as n approaches infinity. When H is acyclic $L_n(H)$ always consists of n copies of H. When H is disconnected, say $H = H_1 + H_2$, we may write $L_n(H) = L_n(H_1) + L_n(H_2)$ where the two parts are generated independently. We shall therefore restrict our attention to H which are connected and contain at least one cycle. We shall look at sentences A in a somewhat stronger language than the usual first order language of graphs.

Definition 12.3.2. *The first order language for lifts over a fixed H consists of equality, adjacency, and unary predicated U_v for each $v \in H$. The statement $U_v(x)$ has the interpretation that vertex $x \in L_n(H)$ lies in the fibre of v or, equivalently, that $x = (v, i)$ for some $1 \le i \le n$. In this case we say x lies* above v.

Example: Let H be a triangle on vertices u, v, w. We may define $L_n(H)$ by selecting $\sigma_1, \sigma_2, \sigma_3$ independently and uniformly from the permutations of $\{1, \dots, n\}$ and letting (u, i) , $\big(v, \sigma_1(i)\big)$ be adjacent and (v, i), $\big(w, \sigma_2(i)\big)$ be adjacent and (w, i), $\big(u, \sigma_3(i)\big)$ be adjacent for $1 \le i \le n$. Consider the event A that there exists no triangle. Then $L_n(H)$ satisfies A if and only if $\sigma_1\sigma_2\sigma_3$ has no fixed point. But $\sigma_1\sigma_2\sigma_3$ is itself a uniformly distributed permutation. By a standard combinatorial exercise the limiting probability of A is then e^{-1}.

Example: Let H be a complete graph on vertices t, u, v, w. The sentence B

$$\neg\Big[\exists_x \exists_y \exists_z \big(U_u(x) \wedge U_v(y) \wedge U_w(z) \wedge x \sim y \wedge y \sim z \wedge z \sim x\big)\Big]$$

may be interpreted that there does not exist a triangle above u, v, w. From the previous example B has limiting probability e^{-1}.

Let T denote the almost sure theory for $L_n(H)$. T contains the following schema. The first statement holds whenever $n \ge r$ and the others hold always.

1. (for all $v \in H$, $r \ge 1$): There exist (at least) r vertices x with $U_v(x)$.
2. (for all $v \in H$)

$$\forall_x \forall_y \Big[\big(U_v(x) \wedge U_v(y)\big) \rightarrow (\neg x \sim y)\Big]$$

   That is, vertices in the same fibre are nonadjacent.
3. (for all nonadjacent $v, w \in H$)

$$\forall_x \forall_y \Big[\big(U_v(x) \wedge U_w(y)\big) \rightarrow \neg(x \sim y)\Big]$$

4. (for all adjacent $v, w \in H$

$$\forall_x U_v(x) \rightarrow [\exists!_y U_w(y) \wedge x \sim y]$$

$$\forall_y U_w(y) \rightarrow [\exists!_x U_v(x) \wedge x \sim y]$$

That is, adjacent fibres are connected by a 1-factor.

Theorem 12.3.1. *For every k and every H almost surely $L_n(H)$ does not contain k vertices with (at least) $k+1$ internal edges.*

Proof: There are $O(n^k)$ choices of k vertices from $L_n(H)$ and $O(1)$ choices for $k+1$ adjacencies. Fix the vertices and adjacencies. Suppose $v_1, \ldots, v_a \in L_n(H)$ and $w_1, \ldots, w_a \in L_n(G)$ lie over $v, w \in H$ respectively and each v_i is required to be adjacent to its corresponding w_i. This occurs with probability $(n-a)!/n! \sim n^{-a}$. In all other cases the adjacencies are not possible. As the 1-factors are independent and the numbers of required adjacencies a add to $k+1$ the probability of having all the desired adjacencies is $O(n^{-(k+1)})$.

What about cycles? By a potential cycle we mean a sequence $v_0, \ldots, v_{s-1} \in H$ with (addition modulo s) all v_i, v_{i+1} adjacent and no $v_i = v_{i+2}$. We do, however, allow repetitions. Thus when H is a triangle on u, v, w this includes the sequences uvw;$uvwuvw$;$uvwuvwuvw$; etc. For each such potential cycle $\mathbf{v} = (v_0, \ldots, v_{s-1})$ let $X_{\mathbf{v}}$ denote the number of cycles $w_0, \ldots, w_{s-1} \in L_n(H)$ (now the w_i are distinct) with w_i lying over v_i.

Theorem 12.3.2. *The $X_{\mathbf{v}}$ are asymptotically independent Poisson distributions of mean one.*

We outline the argument. For a given $\mathbf{v}$ there are $\sim n^s$ potential cycles lying above it and each occurs with probability $\sim n^{-s}$ so that $E[X_{\mathbf{v}}] \sim 1$. Moreover, take any finite family $\mathbf{v_1}, \ldots, \mathbf{v_r}$ or such vectors and positive integers $m_1, \ldots, m_r$. Consider a configuration in $L_n(H)$ consisting of m_i cycles above $\mathbf{v_i}$ for $1 \leq i \leq r$, all disjoint. If t is the number of vertices in such a configuration there will be $\sim n^t$ such potential configurations and each occurs with probability $\sim n^{-t}$ so the expected number of such configurations is asymptotically one. Using Inclusion-Exclusion it then follows from general results that the $X_{\mathbf{v}}$ are asymptotically independent Poissons of mean one.

The above bounds on the number of cycles of a given type allow us to show that almost surely there are many vertices of $L_n(H)$ whose local neighborhoods are trees.

Theorem 12.3.3. *Let graph H, vertex $v \in H$, and positive integers r, d be fixed. Then the following holds almost always in $L_n(H)$: For all $x_1, \ldots, x_r \in L_n(H)$ there exists $x \in L_n(H)$ with x above v such that the d-neighborhood of v is a tree and is disjoint from the d-neighborhoods of the $x_1, \ldots, x_r$.*

Proof: Let $\epsilon > 0$ be fixed and arbitrarily small. The expected number of cycles of length at most $2d$ is bounded by some constant K [as there is a contribution of one from each cycle type] and so with probability at least $1-\epsilon$ the number of such cycles is less than some constant K_1. But then (as the degrees are bounded) there are less than some constant K_2 vertices within distance $2d$ of such a cycle or within distance $2d$ of one of the x_i. With $n > K_2$ some x lying over v will then neither be near a small cycle nor near one the of x_i and hence will satisfy the conditions.

Theorem 12.3.4. *Let $n_{\mathbf{v}}$ be nonnegative integers for every potential cycle $\mathbf{v}$. To the almost sure theory T for $L_n(H)$ add the first order sentences "There are precisely $n_{\mathbf{v}}$ cycles in $L_n(H)$ lying above $\mathbf{v}$. The extended theory T^+ is complete.*

Proof: Let G_1, G_2 be two models of T^+. Let $S_1 \subset G_1$, $S_2 \subset G_2$ be those vertices lying in cycles of length at most $2d$ where $d = \frac{3^k-1}{2}$. The values $n_{\mathbf{v}}$ specify the number of such cycles and so the restrictions of G_1 to S_1 and G_2 to S_2 are isomorphic. When $x \in G_1 - S_1$ its d-neighborhood has no cycles (as otherwise it would be in S_1). From Theorem 12.3.3 one can find $y \in G_2 - S_2$ with the same d-neighborhood. G_1, G_2 now meet the conditions of Theorem 2.4.4 and so Duplicator wins the Ehrenfeucht Game $\textsc{Ehr}(G_1, G_2; k)$.

Theorem 12.3.5. *Let A be a first order sentence in the lift language of Definition 12.3.2. Then A has a limiting probability which can be expressed as a finite sum of finite products of terms, where each term is either of the form $e^{-1}/k!$ for some integer k or of the form $1 - \sum_{i \le k} e^{-1}/k!$ for some integer k.*

Proof: Let A have quantifier depth k. From Theorem 12.3.4 the truth value of A is almost surely determined by the numbers $n_{\mathbf{v}}$ of cycles for the different potential cycles $\mathbf{v}$. Those $\mathbf{v}$ of length greater than 3^k do not affect the truth value as such large cycles cannot be exploited by Spoiler in the Ehrenfeucht game. All values $n_{\mathbf{v}} > k$ are equivalent as in a k move game Spoiler could only play in k of them. Thus the truth value of A is determined by a finite number of $n_{\mathbf{v}}$, each of which take on the possible values $0, 1, \ldots, k$ and "Many." Theorem 12.3.2 gives each such set of values has a limiting probability which is the product of terms as described.

The transcendence of e allows us to express this result is a form that, at least to this author, epitomizes the strangeness of this Strange Logic.

Theorem 12.3.6. *No first order A can have limiting probability $\frac{1}{2}$ over the random lift $L_n(H)$.*

Bibliography

1. N. Alon and J. Spencer, The Probabilistic Method, 2nd Edition. Wiley. 2000
2. B. Bollobás, Random Graphs, Academic Press, 1985
3. K.J. Compton, 0-1 Laws in Logic and Combinatorics, in *Algorithms and Order*, I. Rival, ed., NATO ASI series, Kluwer Academic Publishers, Dordrecht, 1988, 353–383
4. P. Erdős, Graph Theory and Probability, Canad. J. Math *11* (1959), 34–38
5. P. Erdős and A. Rényi, On the Evolution of Random Graphs, Mat. Kutató Int. Közl *5* (1960), 17–60
6. A. Ehrenfeucht, An application of games to the completeness problem for formalized theories, Fundam. Math. *49* (1961), 129–141
7. R. Fagin, Probabilities in Finite Models, J. Symbolic Logic *41* (1976), 50–58
8. Y.V. Glebskii, D.I. Kogan, M.I. Liagonkii and V.A. Talanov, Range and degree of realizability of formulas in the restricted predicate calculus, Cybernetics *5*, 142–154 (Russian original: Kibernetica *5* (1969), 17–27)
9. S. Janson, T. Łuczak and A. Ruciński, Random Graphs, Wiley, 2000.
10. S. Janson, T. Łuczak and A. Ruciński, An exponential bound for the probability of nonexistence of a specified subgraph in a random graph. In *Random Graphs '87*, Proceedings, Poznań, 1987, eds. M. Karoński, J. Jaworski and A. Ruciński, Wiley, 73–87.
11. J. Kim and V. Vu, Concentration of multivariate polynomials and its applications, Combinatorica *20* (2000), 417–434.
12. T. Łuczak and S. Shelah, Convergence in homogeneous random graphs, Random Structures and Algorithms, *6* (1995), 371–392
13. J. Lynch, Probabilities of first-order sentences about unary functions, Trans. Amer. Math. Soc. *287* (1985), 543–568
14. J. Lynch, Properties of Sentences about Very Sparse Random Graphs, Random Structures and Algorithms *3* (1992), 33–54
15. S. Shelah and J. Spencer, Zero-One Laws for Sparse Random Graphs, J. Amer. Math. Soc. *1* (1988), 97–115
16. S. Shelah and J. Spencer, Random Sparse Unary Predicates, Random Structures & Algorithms *5* (1994), 375–394
17. J. Spencer, Zero-One Laws via the Ehrenfeucht Game, Discrete Appl. Math. *30* (1991) 235–252
18. J. Spencer, Countable Sparse Random Graphs, Random Structures and Algorithms *1* (1990), 205–214
19. J. Spencer, Zero-One Laws with Variable Probability, Journal of Symbolic Logic *58* (1993) 1–14
20. J. Spencer and K. St. John, Random Unary Predicates: Almost Sure Theories and Countable Models, Random Structures & Algorithms *13* (1998), 229–248
21. J. Spencer and G. Tardos, Ups and Downs of First Order Sentences on Random Graphs, Combinatorica *20* (2000), 263–280

22. J. Spencer and L. Thoma, On the limit values of probabilities for the first order properties of graphs, *in* Contemporary Trends in Discrete Mathematics, DIMACS Series vol. 49, Amer. Math. Soc., R. Graham et. al., eds., Amer. Math. Soc. 1999, pp 317–336
23. P. Winkler, Random structures and zero-one laws, *Finite and Infinite Combinatorics in Sets and Logic*, N.W. Sauer, R.E. Woodrow and B. Sands, eds., NATO Advanced Science Institutes Series, Kluwer Academic Publishers, Dordrecht (1993), 399–420.

Index

GPSR Compliance

The European Union's (EU) General Product Safety Regulation (GPSR) is a set of rules that requires consumer products to be safe and our obligations to ensure this.

If you have any concerns about our products, you can contact us on ProductSafety@springernature.com

In case Publisher is established outside the EU, the EU authorized representative is:

Springer Nature Customer Service Center GmbH
Europaplatz 3
69115 Heidelberg, Germany

Batch number: 10584448

Printed by Printforce, the Netherlands